W0017857

CFD Module
Turbulent Flow Modeling

LICENSE, DISCLAIMER OF LIABILITY, AND LIMITED WARRANTY

By purchasing or using this book (the "Work"), you agree that this license grants permission to use the contents contained herein, but does not give you the right of ownership to any of the textual content in the book or ownership to any of the information or products contained in it. *This license does not permit uploading of the Work onto the Internet or on a network (of any kind) without the written consent of the Publisher.* Duplication or dissemination of any text, code, simulations, images, etc. contained herein is limited to and subject to licensing terms for the respective products, and permission must be obtained from the Publisher or the owner of the content, etc., in order to reproduce or network any portion of the textual material (in any media) that is contained in the Work.

MERCURY LEARNING AND INFORMATION ("MLI" or "the Publisher") and anyone involved in the creation, writing, or production of the companion disc, accompanying algorithms, code, or computer programs ("the software"), and any accompanying Web site or software of the Work, cannot and do not warrant the performance or results that might be obtained by using the contents of the Work. The author, developers, and the Publisher have used their best efforts to insure the accuracy and functionality of the textual material and/or programs contained in this package; we, however, make no warranty of any kind, express or implied, regarding the performance of these contents or programs. The Work is sold "as is" without warranty (except for defective materials used in manufacturing the book or due to faulty workmanship).

The author, developers, and the publisher of any accompanying content, and anyone involved in the composition, production, and manufacturing of this work will not be liable for damages of any kind arising out of the use of (or the inability to use) the algorithms, source code, computer programs, or textual material contained in this publication. This includes, but is not limited to, loss of revenue or profit, or other incidental, physical, or consequential damages arising out of the use of this Work.

The sole remedy in the event of a claim of any kind is expressly limited to replacement of the book, and only at the discretion of the Publisher. The use of "implied warranty" and certain "exclusions" vary from state to state, and might not apply to the purchaser of this product.

CFD MODULE
Turbulent Flow Modeling

Mehrzad Tabatabaian, PhD, PEng

MERCURY LEARNING AND INFORMATION

Dulles, Virginia
Boston, Massachusetts
New Delhi

Copyright ©2015 by MERCURY LEARNING AND INFORMATION LLC. All rights reserved.

This publication, portions of it, or any accompanying software may not be reproduced in any way, stored in a retrieval system of any type, or transmitted by any means, media, electronic display or mechanical display, including, but not limited to, photocopy, recording, Internet postings, or scanning, without prior permission in writing from the publisher.

Publisher: David Pallai
MERCURY LEARNING AND INFORMATION
22841 Quicksilver Drive
Dulles, VA 20166
info@merclearning.com
www.merclearning.com
(800) 232-0223

M. Tabatabaian. *CFD Module: Turbulent Flow Modeling*
ISBN: 978-1-937585-43-3

Generated images were made using COMSOL *Multiphysics*® and are provided courtesy of COMSOL.

Figures 1.1 through 1.5 in Chapter 1 were drawn by Jean Leary Pallai and were adapted from photographs appearing in M. Van Dyke, An Album of Fluid Motion, Stanford, CA: The Parabolic Press, 1982.

The publisher recognizes and respects all marks used by companies, manufacturers, and developers as a means to distinguish their products. All brand names and product names mentioned in this book are trademarks or service marks of their respective companies. Any omission or misuse (of any kind) of service marks or trademarks, etc. is not an attempt to infringe on the property of others.

Library of Congress Control Number: 2015934496

151617321   This book is printed on acid-free paper.

Our titles are available for adoption, license, or bulk purchase by institutions, corporations, etc. For additional information, please contact the Customer Service Dept. at 800-232-0223(toll free).

All of our titles are available in digital format at authorcloudware.com and other digital vendors. Companion files (figures and code listings) for this title are available by contacting info@merclearning.com. The sole obligation of MERCURY LEARNING AND INFORMATION to the purchaser is to replace the disc, based on defective materials or faulty workmanship, but not based on the operation or functionality of the product.

To my wife
for her support, patience, and encouragement

CONTENTS

PREFACE

Advances in computational methods, software tools and computer power have made engineering and scientific computations more available and economically viable in recent decades. Modeling has become a mainstream, if not a *necessary,* step in engineering analysis and design of products, processes, and systems. However, the vast and powerful facilities available in current and modern computational software packages do not sufficiently match the required training that engineering and science students often receive. Therefore, they may not have all the background training required to use software packages. This has created a challenge for industry, to have trained professionals who can create "reliable" models and fully utilize commercially available software packages. On the other hand, students, engineers, and scientists may not have the luxury of time and training to learn all the necessary technical subjects like physics, mathematical modeling, numerical methods, and programming languages. One of these challenging topics is turbulence modeling. In industry, there is increasingly a pressing need for fluid flow computations that involve turbulence modeling. This book aims to help fill the gap on the topic of turbulence modeling in general, as well as applications of COMSOL®- a commercially multiphysics software tool with turbulence modeling facilities.

This book is written for practitioners in scientific and engineering fields e.g., fluid mechanics, biomechanics, and other related fields, as well as students in engineering. The main objective of the material is to introduce the fundamental physics, characteristics, and common models of turbulent flows for practical and industry applications using COMSOL. Industrial and engineering discussion questions and examples are presented with their solutions that could serve as a guide for modeling similar or more complicated turbulent flow problems.

Turbulent flows are complex; hence reliable and "accurate" models are challenging, and in many cases, require parameters that are specific to each problem and relevant to its boundary and initial conditions. Developing comprehensive models for industrial and practical applications has been a subject of research and development since Boussinesq [1], who proposed the concept of *eddy viscosity* in 1877, and Osborne Reynolds [2] who proposed in 1895, the so-called Reynolds decomposition method and developed RANS (Reynolds-Averaged Navier-Stokes) equations. Although some historical research [3] indicates that Boussinesq attempted and partially derived simultaneously, in his 1877 paper, averaged equations for turbulence and the concept of eddy viscosity. Since then, more modern numerical methods and approaches (e.g., LES[1], DNS[2], versions of RANS) have been developed and mainstreamed by researchers with commonly available computer resources for turbulence modeling. These models cover a wide range, both in terms of their complexity and the computer resources required. In this book we cover turbulence models that are widely used and are readily available for applications in the CFD[3] module of COM-SOL, versions 4.4 and 5.

As mentioned, our objective is to introduce the topic of turbulence modeling, specifically fundamental physics and characteristics of turbulence, that are essential for a modeler. In addition, a collection of examples and modeling guidelines through which readers can build their own models is provided. We hope that practitioners, who need to perform turbulence modeling for their related industries, will find this book useful to understand turbulence modeling fundamentals and facilitative for easier navigation of modeling tools available in COMSOL.

A *flexible-level* approach is used to present the materials along with practical examples. Mathematical fundamentals and models, relevant physics, and engineering principles are integrated into discussion questions while the turbulent flow models fundamentals are described in Chapter 2. References which contain more in-depth physics, technical information, analysis, derivations of governing equations (see the Appendix), and data are referred to throughout the book. This approach allows readers to learn the materials at their desired level of complexity.

CFD Modeling: Turbulent Flow Modeling could be used as a textbook, in academia, or as a reference for practitioners in indus-

---

[1.] Large Eddy Simulation
[2.] Direct Numerical Simulation
[3.] Computational Fluid Dynamics

try. Examples provided in this book can be considered as "lessons" through which background physics could be explained in more detail. Exercise problems, or their variations, could be used as homework assignments, as well.

We use the COMSOL software tool (versions 4.4)[4] for solving the examples. *However, readers may upload or build all of the solved examples using version 5.* We also use the new facilities in version 5, like *Application Builder*, for some of the examples presented. Where and when suitable, we also compare the modeling results with existing analytical, experimental, or other relevant models. Detailed step-by-step instructions are provided to build the relevant model for each example. However, it is recommended that readers, specifically students, go through all models to master applications of COMSOL turbulent modeling tools. Readers are assumed to possess a base of knowledge on the topic of fluid dynamics, calculus, differential equations and principles of numerical modeling used in CFD (e.g., finite element methods).

This book is composed of the following four chapters and one appendix:

Chapter 1: Turbulence Fundamentals – A Summary

In this chapter we give some examples of industrial and engineering turbulent flows and discuss important observations drawn from these flows and conclude main properties and characteristics of turbulence. We also mention main turbulent flow types and relevant length and time scales associated with turbulent flows. Some exercise problems are given at the end of this chapter.

Chapter 2: Turbulence Modeling – Approaches and Models

In this chapter we provide an overview of modeling approaches for turbulent flows, from common to more advanced models. Main assumptions and hypotheses are discussed and among all available models, we focus on RANS type of models including those available in COMSOL. The *Closure* problem and relevant solutions for RANS equations are discussed along with corresponding governing equations for exact turbulent kinetic energy and its dissipation. Detailed

---

[4] *http://www.comsol.com/support/releasehistory/*

discussions for RANS turbulent models, available in COMSOL, and their corresponding model equations are given. Finally, a general guideline for turbulent models applications is provided.

Chapter 3: COMSOL Multiphysics – Overview and CFD Module

In this chapter we introduce the main features and structure of COMSOL (versions 4.4 and 5), including the CFD module and models for turbulence. A worked-out example is presented along with instructions to build the corresponding model and building a model application using COMSOL *Application Builder*. Also, a general guideline for building a model in COMSOL is provided.

Chapter 4: Turbulent Flow Models – Application Examples

In this chapter we use COMSOL to model examples of flow problems involving turbulence. Several examples and step-by-step instructions are presented for building these models in COMSOL 4.4 (or equivalently in version 5), in addition, to *Application Builder* applications (available in version 5), and interpretation of results. Where applicable, modeling results are compared against relevant existing models or experimental results. Readers will benefit from modeling examples provided in this chapter if the preceding chapters' materials are covered beforehand. The *models are available on the companion disc* and can be used either as independent examples or as a starting point for building similar models with modifications.

Appendix: Derivation of Governing Equations

In this appendix we present detailed derivations of the governing equations for Reynolds stress, turbulent kinetic energy, and turbulent kinetic energy dissipation. The exact governing equations are derived. Readers may refer to these derivations for more in-depth realizations of the terms involved and their physical interpretations. However, the main chapters are presented such that the continuity of the materials is preserved. We recommend that users first become familiar with the main topics and modelling techniques presented in the main chapters of this book and then read through this appendix for further detail, in addition to casual references to these derivations. We also believe that having all the details of derivations for the main governing equations relevant to turbulence modeling, as a collection, would be a valuable resource to the readers.

Mehrzad Tabatabaian, Ph.D., P. Eng.
Vancouver, BC
April 2015

1

TURBULENCE FUNDAMENTALS— A SUMMARY

OVERVIEW

It is challenging to provide a universally acceptable and precise definition for turbulence. However, despite their complexity, turbulent flows have 'common' properties or characteristics which separate them from laminar ones. More common in industry and nature than laminar flows, i.e. omnipresent, turbulent flows are easy to recognize. Water flowing in a river, milk mixing in a cup of tea by stirring, smoke coming out of a stack, pipe flow, the air-fuel mixture in an engine cylinder, and blood flow in main arteries are examples of turbulent flows.

It is a common exercise in the development of science that one observes a phenomenon and then tries to develop a theory or model for it. Alternatively, one could propose a hypothesis, for a phenomenon, and validate it by performing experiments. In the case of turbulence both approaches have been employed by researchers, mainly due to the complexity of this phenomenon-which still remains an 'unsolved' problem of classical physics [4], [5]. We start our discussion by listing some observations we might have from visual examination of some turbulent flow illustrations and pictures. We hope that this helps to answer questions like *What is turbulence?*, *Where does it come from?*, and *Why study turbulence?* We then use our observations, collectively, to list some of the characteristics of turbulent flows.

SOME OBSERVATIONS OF TURBULENT FLOWS

Van Dyke [6] assembled a collection[5] of pictures or flow visualizations of turbulent flows either made in laboratory experiments or naturally occurring. The National Committee for Fluid Mechanics Films (NCFMF) published a series of films [7], which includes one episode on turbulence [8]. We encourage readers to watch these films which discuss turbulence and related topics, like vorticity and flow instabilities. Readers can find newer, mostly color pictures and videos from the Division of Fluid Mechanics of APS [9]. Another collection, *A Gallery of Fluid Motion* [10], has a section on turbulence. Also, the John Hopkins Turbulence database [11] is an excellent source for turbulence related topics (*www.turbulence.pha.jhu.edu*). Besides these sources, modern flow visualization and measurement techniques (e.g. particle imaging velocimetry, laser Doppler velocimetry, and multi-sensor hot-wire probe) can be used to bring more insights into turbulent flow problem and provide more pieces of information for this complex topic.

We use the Van Dyke album for our discussions, here. Figures 1.1 through 1.5 are adaptations of photographs in the album. Figure 1.1 shows a laminar stream of air flows from a circular tube into ambient. The stream is made visible by a smoke wire.

The following observations could be made:

- Close to the nozzle, flow becomes unstable and creates vortices at the boundary while the flow at the inner region seems more stable and 'laminar.'

- The velocity gradient is very large at the boundary of the jet, i.e. a shear layer. The edge of the jet develops axisymmetric oscillations.

---

[5] Henceforth referred to as the Van Dyke album.

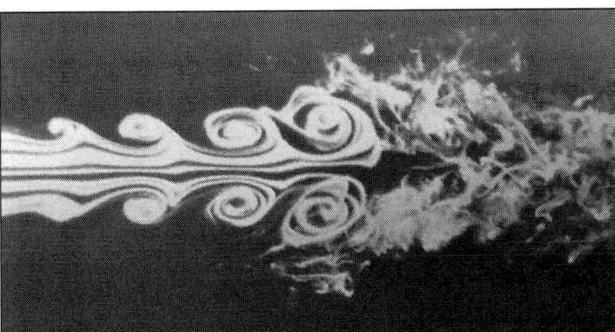

FIGURE 1.1: A laminar stream of air flows from a circular tube at Re = 10^4. (Adapted from picture 102 of the Van Dyke album; photograph by Robert Drubka and Hassan Nagib.)

- It seems that instabilities at the boundary dissipate into the inner region, or inner flow is entrapped into the turbulence developed at the jet boundary.

- The jet flow turns into large vortices and 'abruptly' all the flow becomes turbulent.

- The downstream turbulent flow seems to have smaller length scale than the vortices upstream.

Readers are encouraged to add their own observations to this list.

Figure 1.2 shows laminar flow of carbon dioxide flowing into ambient air at a speed of 127 ft/s (about 38.71 m/s) from a 0.25 inch (6.35 mm) nozzle.

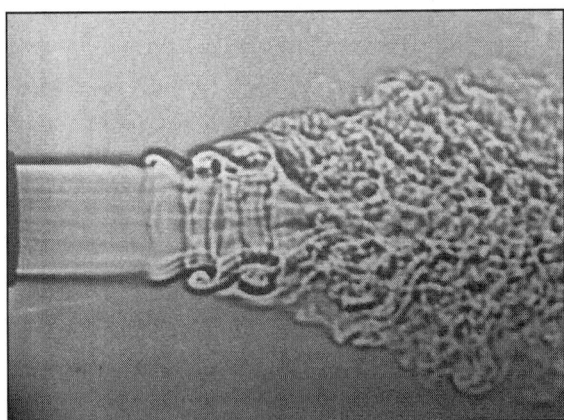

FIGURE 1.2: A laminar stream of CO_2 flows into air from a circular tube at Re = 3×10^4. (Adapted from picture 117 of the Van Dyke album; photograph by Fred Landis and Ascher H. Shapiro.)

The following observations could be made:

- At a distance of about ¼ inch (the diameter of the nozzle) flow becomes unstable and creates vortices at the boundary while the flow at the inner region seems more stable.

- Vortices expand at a smaller distance as compared to the jet in Figure 1.1.

- The velocity gradient is very large at the boundary of the jet, i.e. a shear layer.

- It seems that instabilities at the boundary dissipate into the inner region, or inner flow is entrapped into the turbulence developed at the jet boundary.

- The jet flow turns into large vortices and 'abruptly' all the flow becomes turbulent.

- The downstream turbulent flow seems to have smaller length scale than the vortices upstream.

From the comparison of these jet flows, shown in Figures 1.1 and 1.2, it seems that at a higher Reynolds number the regions of instability shrink and turbulence happens at a point which is closer to the exit, relatively. These two flows are called free jet flows.

In Figure 1.3, a turbulent boundary layer develops on a 3.3 m long plate located in a wind tunnel. The Reynolds number, Re, is 3500 based on momentum thickness (proportional to $1/\sqrt{Re}$). The visualization of the flow is made possible by a smoke wire located near the leading edge of the plate and illuminated by a vertical slice of light.

The following observations could be made:

- The flow seems turbulent, but intermittent, close to the plate boundary.

- Turbulence bursts closer to the plate boundary seem to have smaller length scale than those away from it.

- The turbulent boundary layer has a finite thickness, as shown in the picture frame, and the main flow remains laminar or stable.

FIGURE 1.3: A turbulent boundary layer over a flat pate at Re=3500. (Adapted from picture 157 of the Van Dyke album; photograph by Thomas Corke, Y. Guezennec, and Hassan Nagib.)

Figure 1.4 shows the flow past a circular cylinder at Re = 2000.

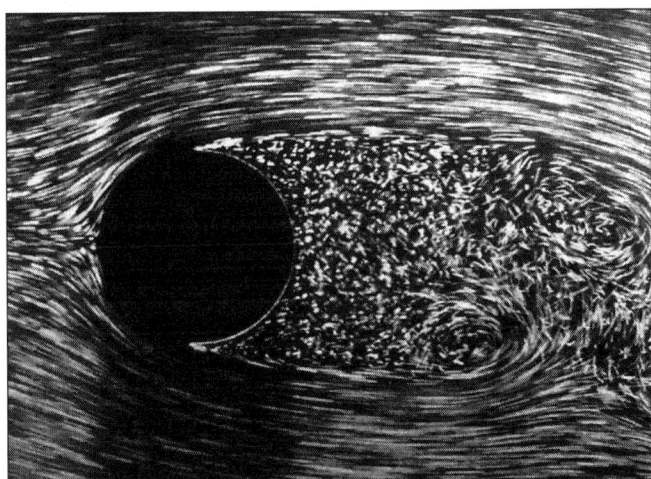

FIGURE 1.4: Trailing vortices and turbulent wake of a cylinder, Re = 2000 (adapted from picture 47 of the Van Dyke album, photograph by Werle and Gallon).

The following observations can be made:

- The boundary layer is laminar over the front (left) side of the cylinder and separates from the cylinder towards the downstream side and creates turbulence in the wake region.

- The velocity gradient is very large at the boundary of the wake and the main flow.

- A pair of vortices is generated at about one diameter distance into the wake.

- Vortices draw the fluid back into the wake, i.e. they rotate in opposite directions.

These types of turbulent flows are called 'bounded' flow, since they interact with a 'solid' wall of an object.

The last picture that we consider for observation in this chapter is a repetition of a famous experiment by Osborne Reynolds, done by N. H. Johannesen and C. Lowe, as shown in Figure 1.5.

The flow regime changes from laminar (shown at the top of Figure 1.5) through transition and finally a fully turbulent flow (shown at the bottom of Figure 1.5). It is interesting to note that vibration caused by modern traffic in the streets close to Manchester University caused the onset of turbulence at a Reynolds number lower than 13000, which was the number found by Reynolds in 1883 in his experiment at the same university/location.

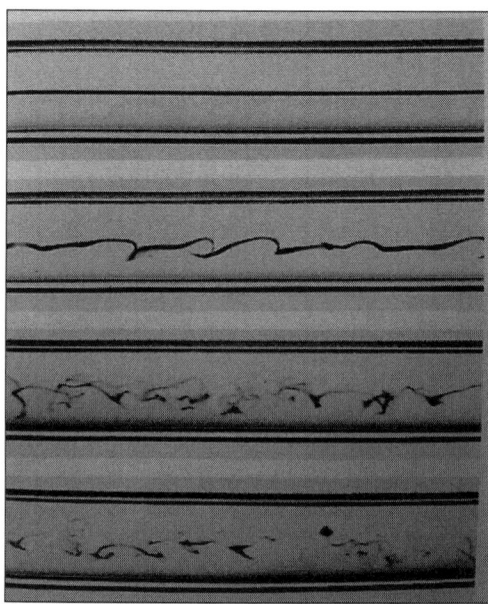

FIGURE 1.5: Repetition of Reynolds' dye experiment, critical Reynolds number about 13000 (adapted from picture 103 of the Van Dyke album).

There are more pictures in the Van Dyke album which could be analyzed and used for discussion and examination. We encourage readers to do this as exercises, in order to familiarize themselves with features and characteristics of turbulent flows.

TURBULENCE CHARACTERISTICS

Reynolds number, $R_e = \dfrac{UL}{\nu}$, is the dimensionless number that characterizes flow regimes, where U and L are the typical velocity and length scales involved, respectively and ν is the kinematic viscosity of the fluid. Reynolds number, generally, represents the ratio of inertia force to viscous forces in a flow. Another useful physical interpretation for Reynolds number is the ratio of viscous diffusion time scale over time scale related to motion of large eddies. When Reynolds number becomes large (as compared to unity) the flow becomes unstable and goes through a transition until a fully-turbulent state is reached. Typical values of Reynolds number depend on the specific geometry and boundary conditions; for example for a typical pipe flow Reynolds number is about 2000–2300 [12]. A laminar flow becomes turbulent by going through a transition. The transition state is very complex but may last a relatively short period of time. Usually, and mainly for many practical and engineering applications, we are interested in the final turbulent state of the fluid flow or the so-called fully developed turbulent flow.

Following our observations from the turbulent flow visualizations mentioned in the previous section, and for a 'better' understanding of turbulence, we list some of its main characteristics or properties [13] along with relevant discussions. These properties are generally accepted (although some arguments exist [14]) in the literature and turbulence community among researchers.

Disorder: Turbulent flows are disorderly or irregular. Fluid particles move around chaotically, apparently 'randomly.' This property of turbulence is so central to its definition that no matter how carefully one tries to reproduce the boundary conditions involved, the flow will never reproduce itself with the same details. In other words, turbulence is very sensitive to initial and boundary conditions, as if it has a 'memory.' This property of turbulence led Osborne Reynolds, and others, to assume that turbulence is a random phenomenon and used statistical methods for analyzing turbulent flows. It should be noted that not all irregular flows are turbulent. For example, looking at the calm ocean surface water waves from the coast, they may appear irregular but are not turbulent.

Diffusive: Turbulent flows are diffusive. This property of turbulence is very important from a practical application point of view and hence useful for practical applications. Diffusion in a stagnant fluid occurs by molecular diffusion effect. Turbulence makes the diffusion or mixing process of quantities much more rapid by making inhomogeneity which might be present in a flow more susceptible to the effect of molecular diffusion [16]. Turbulence diffusion enhances the mixing of momentum, mass, energy and any other inhomogeneity present in a fluid flow. A very simple experiment readers may want to try is to add a drop of ink in a stagnant volume of water in a glass. The ink will diffuse throughout the water and it takes a while until the ink concentration becomes uniform throughout the volume. Now if you stir the water in the glass, diffusion happens much faster due to turbulence and until ink uniform concentration occurs.

Dissipative: Turbulent flows are dissipative, which means that they 'transform' energy (i.e., energy is transferred and finally dissipated). This is one of the curious properties of turbulence. As mentioned previously, Reynolds number determines if a flow is turbulent or not, but when turbulence occurs, then the Reynolds number has very little importance, as far as large scale motion or motion of large eddies is concerned (for example recall the Moody diagram [12]). However, the larger the Reynolds number, the finer the small scale eddies become. To understand this better, we discuss the concept of 'energy cascade' [18], as one of the suggested mechanisms [19] for turbulence dissipation. In a turbulent flow some of the energy in the main flow is converted to turbulent energy through formation of large eddies or vortices. Large eddies break down into smaller, and yet smaller, ones which carry the turbulent energy with themselves. Large eddies carry small eddies with themselves, as well, i.e. they overlap in the space. The break-down process continues until the energy can be dissipated through viscosity into heat. However, it has been suggested that energy from small eddies might be transferred

to larger ones, at least locally [19]. The length scale of the smallest

eddies is defined by the Kolmogorov length scale, $\eta = \left(\dfrac{v^3}{\varepsilon}\right)^{1/4}$ (with

associated time scale $\tau = \left(\dfrac{v}{\varepsilon}\right)^{1/2}$ and velocity scale $\vartheta = (\varepsilon v)^{1/4}$), where

v is kinematic viscosity of the fluid, and ε is average turbulence energy dissipation per unit time per unit mass. Notice that for these representative variables the Reynolds number Re $= \vartheta\eta/v$ is unity. For example; for a car moving at a speed of about 105 km/h, $\eta \approx 1.8 \times 10^{-4}$ m [20]. In other words the energy cascade break down process stops at length scales of order η, since the energy has been broken down to a level that can be converted to heat (or internal energy) by viscosity. One of the key assumptions, usually made, is the 'continuity' of energy cascaded through eddies. In other words the energy dissipated (per unit time) at the Kolmogorov scales to heat is equal, with very good approximation, to the one that is dissipated from larger eddies. It can be shown [21] that the ratio of velocities, times, and lengths at a typical large eddy to those of Kolmogorov's are $Re^{1/4}$, $Re^{2/4}$, $Re^{3/4}$, respectively. Also due to the local isotropy assumption for turbulence, at very small length scale the structure of small eddies are very much similar from one turbulent flow to another. In other words, the small eddies structure in turbulent flows of a jet stream and a channel flow, for example, are very similar [8].

Continuum: Turbulence is a time dependent three dimensional continuum phenomenon [20]. Navier-Stokes (N-S) equations govern turbulent flows at every space point in the domain of the flow. However, when time averages of flow quantities are considered, we can consider the turbulent flow as two-dimensional, where applicable. No matter how small the eddies become in size, due to turbulent dissipation, still they are much bigger than the molecular length scale of the fluid. In other words, we assume that for turbulent flows the Knudsen number (i.e. ratio of mean free path of fluid molecules over the small eddies length scale) is very small (as compared to unity) while the Reynolds number is very large. This property of turbulence is useful for engineering and industry applications since we can, in principle, solve N-S equations for turbulent flows. This is the principle of DNS method.

Multiple length and time scales: Turbulent flows have a wide range of length and time scales, and hence many degrees of freedom. It is this feature of turbulence that makes it difficult to 'solve' the turbulence problem, both theoretically and numerically as well as experimentally. If data are taken, for example velocity, with a probe in a turbulent flow and analyzed, using Fourier analysis (i.e. mapping time domain to frequency domain), we can obtain the turbulent energy spectrum $E(k)$ versus wave number k (or inverse of the associated length scales) of the signals [22], as shown schematically in Figure 1.6.

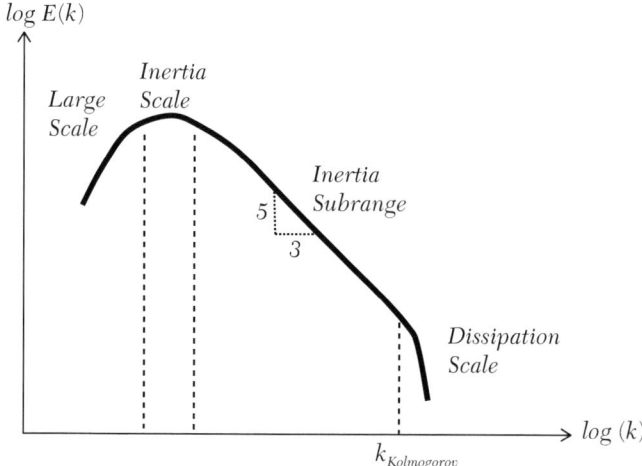

FIGURE 1.6: Turbulent energy spectrum E(k) versus wave number k, log-log scales.

As seen in this figure, most of the energy is generated at large length scale eddies, which correspond to small wave lengths. The energy starts to dissipate at a slope of –5/3 in the so-called inertial range and at a higher slope in the viscous dissipation range. The slope –5/3 was predicted by Kolmogorov (or the Kolmogorov spectrum law) in 1941 [23], part of so called 'K41 theory' [24]. The ratio of the length scales corresponding to maximum energy and that of dissipation increases with Reynolds number, as $Re^{3/4}$ [24]. Therefore for high-Reynolds-number turbulent flows there are many degrees of freedom.

Source: Turbulence could be initiated in a shear flow and/or by buoyancy force. When shear stress is large the flow becomes unstable and vortices are generated. The most important source for flow instabilities and hence, turbulence is believed to be the velocity gradient[6] $\nabla u = \dfrac{\partial u_i}{\partial x_j} = u_{i,j}$. This quantity is actually a tensor with nine components, in general, in three dimensional space. When the velocity gradient becomes very large, for example in a shear flow, extra strain and rotation are created in the flow. In other words, a material element of the fluid is strained and rotates. This can be mathematically shown, as follow. We decompose the velocity gradient into a symmetric strain tensor and an asymmetric rotation tensor:

$$\frac{\partial u_i}{\partial x_j} = \frac{1}{2}(u_{i,j} + u_{j,i}) + \frac{1}{2}(u_{i,j} - u_{j,i}) \; ; \; i \text{ and } j = 1,2,3 \qquad (1.1)$$

$$= \frac{1}{2}\begin{bmatrix} 2u_{1,1} & u_{1,2}+u_{2,1} & u_{1,3}+u_{3,1} \\ u_{2,1}+u_{1,2} & 2u_{2,2} & u_{2,3}+u_{3,2} \\ u_{3,1}+u_{1,3} & u_{3,2}+u_{2,3} & 2u_{3,3} \end{bmatrix}$$

$$+ \frac{1}{2}\begin{bmatrix} 0 & u_{1,2}-u_{2,1} & u_{1,3}-u_{3,1} \\ u_{2,1}-u_{1,2} & 0 & u_{2,3}-u_{3,2} \\ u_{3,1}-u_{1,3} & u_{3,2}-u_{2,3} & 0 \end{bmatrix}$$

Recall that the rate of strain tensor is $e_{ij} = \dfrac{1}{2}(u_{i,j} + u_{j,i})$, and rotation is $\xi_{ij} = \dfrac{1}{2}(u_{i,j} - u_{j,i}) = -\dfrac{1}{2}\mathcal{E}_{ijk}\Omega_k$, where vorticity is defined as the curl of velocity, $\Omega_i = \mathcal{E}_{ijk}u_{k,j}$, and $\mathcal{E}_{ijk}$ is the permutation symbol (or Levi-Civita tensor). The mathematical relations given here simply express that a fluid element undergoes a net solid-body-like translation, a net solid-body-like rotation (with angular velocity of $\dfrac{1}{2}|\Omega_k|$), and a pure strain. As mentioned previously, according to the energy cascade scenario, the vortices (or swirls) are then stretched and broken

---

[6] We use index or tensor notation, where summation convention applies. Please see [41]

down to smaller ones. A vorticity vector is associated with every point in a fluid. These vectors are tangent to the vortex line. Accordingly, and simplistically, turbulent flows can be thought of as vortex lines which are entangled like 'spaghettis' which carry energy with themselves as their resolutions increase.

The list given above is not exclusive, but contains the important and main properties of turbulence. For further readings, readers may consult Batchelor [25], Hinze [26], Townsend [27], Tsinober [14], Cebeci [13], Tennekes and Lumley [22], and Wilcox [20], among many other resources available.

So far we have made some observations using pictures of turbulent flows and had some discussions about their characteristics. In the next section we briefly explain and list some turbulent flow types.

TURBULENCE FLOW TYPES

In general turbulent flows can be categorized into three types; free, semi-confined, and fully confined. Examples of free turbulent flows are jet, wake, and free shear flow. Examples of semi-confined turbulent flows are; turbulent boundary layer over a flat plate, flow passing over an airfoil, wind, water flow in a river, and flow passing a sharp solid edge. Examples of fully confined flows are; flow in a pipe, blood flow in a human heart, and flow inside an internal combustion engine cylinder. Here we include fluids with different densities or viscosities; stratified flows for example. Characteristics associated with each flow type should be studied and identified before choosing a turbulent model. For a free shear flow, shown in Figure 1.2, there exists an expanding region with high velocity gradient. The velocity gradient decreases as fluid flows downstream. For a wake behind a body, as shown in Figure 1.4 for a cylinder, there exists a region near the trailing edge of the body with decreasing or even reverse velocity and vortices with high velocity gradient at the boundary with the main flow. For the cases of semi-confined and fully-confined flows, the turbulent boundary layer develops from the solid wall and expands into the main flow while velocity gradient is relatively high close to the boundary. For these flows, separation of boundary layer may occur, which creates a challenge for modeling. Schematics of some turbulent flows are shown in Figure 1.7.

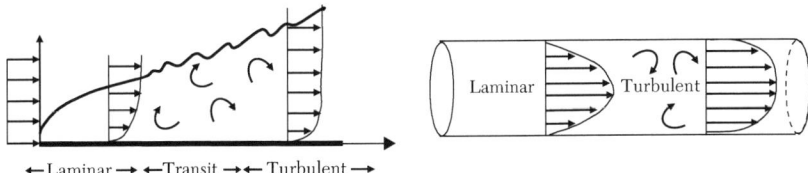

FIGURE 1.7: Schematic of semi-confined (left) and fully-confined (right) turbulent flows.

Readers are referred to, and are encouraged to study behavior of different types of turbulent flows using sources and references mentioned in this chapter. Understanding the flow behavior along with the materials covered in Chapter 2 can help users with choosing the 'right' turbulent model for the problem at hand.

EXERCISES

1.1. List your observations from pictures 152, 153, and 167 from Van Dyke.

1.2. Describe two sources of turbulence.

1.3. Describe different turbulent flow types.

1.4. List three characteristics of turbulence and describe them.

1.5. Intermittency is observed and also considered as one of the characteristics of turbulence. Investigate and describe this property of turbulence.

1.6. Perform research work on the claim that turbulence is a random phenomenon by identifying properties of randomness and comparing to properties of turbulence. Based on your research conclude whether turbulence is a random phenomenon or not.

1.7. Leonardo da Vinci did some observations of turbulent water flow and made some descriptions of it: "... *the smallest eddies are almost numberless, and large things are rotated only by large eddies and not by small ones, and small things are turned by small eddies and large.*" Analyze this quote against turbulence characteristics mentioned in this chapter.

1.8. C. Fukushima and J. Westerweel [28] (at the Technical University of Delft, The Netherlands) made a flow visualization of a turbulent jet. The flow was made by laser-induced fluorescence. The jet exhibited a wide range of length scales. Study this flow and list its characteristics by observation.

CHAPTER 2

TURBULENCE MODELING– APPROACHES AND MODELS

OVERVIEW

Modeling of turbulence progress is very much interwoven with its R&D history. As previously mentioned, observation, theoretical, experimental, and numerical work done so far with the goal of solving this challenging problem is extensive and many researchers, engineers, and scientists in the last two centuries have been involved and contributed to this field and many more continue to make advances.

We summarize the history of these developments, basically from their approaches point of view on turbulence modeling and then we move on to explain the modeling methods with emphasis on models available in COMSOL. For those interested, we encourage the exploration of further readings on the history of turbulence and its development in, for example, Chapman and Tobak [29], Lumley and Yaglom [30], McDonough [19], and a more recent publication on the history of turbulence and its development, *A Voyage Through Turbulence* [31] including a related video series made available by the European Turbulence Conference 2011 (*http://etc13.fuw.edu. pl/historical-turbulence*).

Chapman and Tobak divided the history of work done on turbulence, starting from the time of Osborne Reynolds through to the mid-1980s, into three movements or approaches: (1) *Statistical*, (2) *Deterministic*, and (3) *Structural (coherent)*. In their paper [29] they listed main contributors to each movement and McDonough [19] extended this list further to about 2000. Our focus here is not extensive coverage of the history but rather capturing the work done on turbulence and its effects on turbulence modeling, as follows:

1. The *Statistical* approach assumes that turbulence is a random phenomenon and hence one cannot capture its details. But statistical methods could be used to capture main features or some 'averaged' behavior of turbulence. This approach, although criticized by some researchers [19], forms the foundation of turbulence modeling for practical and/or engineering applications. The pioneer researcher for this movement was Osborne Reynolds [2]. Practical and engineering applications of turbulence modeling follow this movement, specifically the methods based on RANS. Several modeling methods are available in COMSOL using the RANS approach. We should also mention here that the work of Reynolds on turbulence came after a fundamental hypothesis proposed by Boussinesq, which required an approximation for eddy viscosity [1], [3] or the so-called 'closure problem.' We will discuss these topics in later sections.

2. The *Deterministic* approach assumes turbulence as chaotic, and seemingly random (although not random by definition) which could be deterministically modeled and predicted using the N-S equations. The pioneering researchers for this movement (directly and indirectly) were Poincare [32], Lorenz [33], and Orszag and Patterson [34]. The latter researchers did the first DNS (Direct Numerical Simulation) type modeling of turbulence. The DNS method requires a huge amount of computer power (number of elements/cells $\mathcal{O}(Re^{9/4})$ and CPU time $\mathcal{O}(Re^3)$), which is becoming gradually available, although not yet commercially, with advances in computer technology and post-processing techniques for visualization and analysis of huge amount of data generated. This method is not, currently, available in COMSOL.

3. The *Structural* approach assumes that turbulence is not a random phenomenon, but rather there exist coherent structures in turbulent flows. Coherent structure has been defined by McDonough [19] *"Coherent structures are not-very-well-defined behaviors in a turbulent flow, but which are identified as being easy to "see," may or may not be of fairly large scale, and are somewhat persistent."* This movement began with Schubauer and Skramstad [35] and observations of Tollmien–Schlicting waves in 1948. This approach is still pursued by many researchers for detecting and analyzing coherent structures in turbulent flows.

An interesting conceptualization of the main points in each movement, mentioned above, is found in Chapman and Tobak [29] and McDonough [19]; the statistical movement is referred to as being a structureless theory with weak power of conceptualization and prediction. The same authors characterize the structural movement as having produced some structure although still weak in theory.

In addition to the turbulent modeling approaches, briefly mentioned above, work has been done on hybridization of the results of each movement with others, for example LES (Large Eddy Simulation) and DES (Detached Eddy Simulation) methods. In LES the objective is to resolve and capture the dynamics of large eddies while the small-scale eddies are modeled. This method sits somewhere between RANS and DNS. LES is becoming more amenable for practical and engineering applications, both because of its availability in commercial software packages and also for the commercial availability of required computer power. The DES method is more recent and tries to combine LES and RANS to model the smaller-scale turbulence in near-wall regions.

Finally, readers are referred to a relatively complete list of definitions of terminology, methods, and approaches used in turbulence literature provided by McDonough [19].

MODELING METHODS

As already mentioned, one of the challenges with turbulence modeling is its wide range of length and time scales. For example,

Spalart [36] identifies two main challenges posed by a turbulence model: (1) boundary layer development, growth, and eventual separation, and (2) predicting the transfer of momentum after separation. Most models are capable of simulating the former, within models of curvature and pressure gradient, whereas the latter is more achievable with complex models. Currently, there are, mainly in use, three methods for modeling turbulence RANS, LES, and DNS, including their modified versions and variations. The wide range of length and time scales associated with turbulence are fully resolved (i.e. not modeled) or partially resolved and/or modeled within each method. This is, schematically, shown in Figure 2.1.

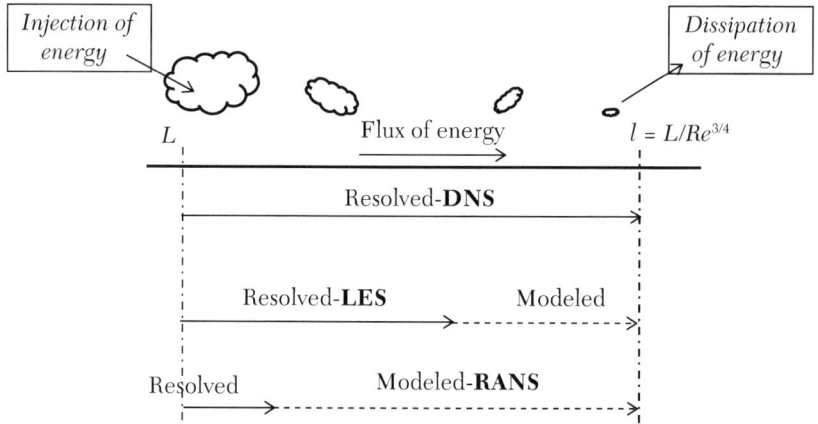

FIGURE 2.1: Extent of turbulence models relative to turbulence eddies and length scales.

As shown, for the RANS method the whole range of energy fluxes and hence corresponding length scales are modeled. This is a huge constraint on turbulence and as a consequence the final results show averaged flow and its mean quantities. Although this is restrictive from the physics of turbulence point of view and its fine-resolution eddies detail, especially at smaller length scales for dissipative eddies, it matches the engineering objective for modeling turbulent flows. In many cases, for engineering type applications (notwithstanding exceptions like turbulent combustion [37]), we are interested in mean or averaged quantities of turbulent flows rather than the details of their turbulent fluctuations. This may be one of the reasons why RANS are more popular in engineering applications compared to other methods like LES and DNS and, in addition

RANS are less demanding on computer power as compared to the LES and DNS methods. In terms of computation time RANS require $\mathcal{O}(Re)$, LES is, $\mathcal{O}(Re^2)$, and DNS is $\mathcal{O}(Re^3)$. For details and comparisons of LES and DNS we refer interested readers to [19], [38], and [39]. In this book, we focus on RANS models and their formulations and applications in engineering. The RANS method is currently the most computationally practical method/approach for turbulence modeling.

RANS EQUATIONS

When a fluid is flowing, unconfined or confined, it carries a certain amount of energy (including kinetic and internal energies) with itself. Fluid particles dissipate part of the energy through friction or viscous effect within the body of the fluid or in contact with the walls of the container. When the energy production exceeds the viscous dissipation mechanism capacity then flow becomes unstable and takes on another mechanism, i.e. turbulence, to dissipate the energy to the level which is again 'digestible' for the viscous dissipation mechanism. A source of flow instability could be shear forces exerted on the fluid. To begin our discussion, we start with laminar flow. A fluid cannot resist shear forces (or stresses) when it is stagnant. A fluid should flow in order to resist shear force. The relationship between shear stress, in a laminar flow, for a viscous fluid and the resulting (time-) rate of strain is given by Newton's law of viscosity, for example for a flow over a plate:

$$\tau = \mu \frac{du}{dy} = \mu \frac{d(X/t)}{dy} = \mu \frac{d}{dt}\left(\frac{dX}{dy}\right) \tag{2.1}$$

where, τ is shear stress component, μ is fluid viscosity, and $u = u(y)$ is the fluid velocity profile in the x-direction. Note that the gradient of velocity (du/dy) is similar to time rate of strain (dX/dy) where X is the relative displacement of the fluid filament under strain. Figure 2.2 shows a typical velocity profile for flow over a solid plate.

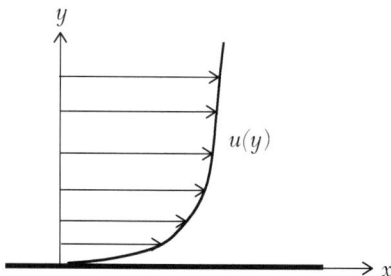

FIGURE 2.2: Typical fluid velocity profile over a plate.

Another force which acts on moving fluid is the inertia force. The balance of the inertia forces and internal forces, both shear and normal forces including pressure-induced force should be in equilibrium with exerted forces on the fluid. The mathematical form of the above-mentioned equilibrium is the well-known conservation of momentum and mass equations [40]. These equations are simply the result of application of Newton's second law and principle of mass conservation for a fluid in motion:

$$\rho\left(\frac{\partial u_i}{\partial t} + u_j u_{i,j}\right) = \sigma_{ij,j} + F_i \tag{2.2}$$

$$\frac{\partial \rho}{\partial t} + (\rho u_i)_{,i} = 0 \tag{2.3}$$

where ρ is fluid density, u_i is velocity vector, σ_{ij} is stress tensor, and F_i is body force. In the above equations summation over repeated indices is applicable ($i =1$, 2, 3 and $j = 1$, 2, 3) and $(.)_{,i}$ denotes differentiation (see [41]). For a Newtonian fluid the constitutive equation for stress tensor is given as $\sigma_{ij} = -p\delta_{ij} + 2\mu\left[e_{ij} - \frac{1}{3}e_{kk}\delta_{ij}\right]$ (see Schlichting [42]). Where δ_{ij} is the Kronecker delta (equals unity for identical value of indices, otherwise is zero), and $e_{ij} = \frac{1}{2}(u_{i,j} + u_{j,i})$ is rate of strain tensor. For an incompressible fluid the continuity equation reduces to $e_{kk} = u_{k,k} = 0$ and stress tensor is given as $\sigma_{ij} = -p\delta_{ij} + 2\mu e_{ij}$. By applying the divergence operation on the stress tensor, the first term on the right-hand side of equation (2.2) reads $\sigma_{ij,j} = -p_{,i} + \mu u_{i,jj}$. The final result, equations (2.4) and (2.5) as given below, are the N-S equations for incompressible fluids, with conservative type body forces absorbed in the pressure term or simply neglected, without loss of generality for our discussion here:

$$\rho\left(\frac{\partial u_i}{\partial t} + u_j u_{i,j}\right) = -p_{,i} + \mu u_{i,jj} \qquad (2.4)$$

$$u_{i,i} = 0 \qquad (2.5)$$

These equations form a system of four PDEs which could be solved, in principle, for four unknowns, i.e. three components of velocity vector and pressure. For derivation and more details see [40]. When we use proper scales for the quantities involved in N-S equations and derive the dimensionless version, we get several dimensionless numbers,[7] including Reynolds number $R_e = \frac{\rho U L}{\mu}$ where, U and L are typical velocity and length scales involved.

N-S equations govern turbulence. However, solution of these non-linear equations for all associated length and time scales that exist in a turbulent flow is a formidable task, both analytically and numerically. Even currently available supercomputers have 'limited' resources for solving these equations for turbulent flows [43]. Osborne Reynolds [2] proposed a solution to this challenge by decomposing turbulent quantities into a mean and a fluctuation part, based on a fundamental 'assumption' that turbulence is a 'random' phenomenon. Therefore quantities like velocity, pressure, density, and heat/material concentration could be decomposed into relevant time-averaged values and their related turbulent fluctuations, as $u_i = \bar{u}_i + u'_i$, $p = \bar{p} + p'$, $\rho = \bar{\rho} + \rho'$, $\Theta = \bar{\Theta} + \Theta'$, respectively. This method is called Reynolds decomposition. For our discussion here, we use velocity as the representative quantity, without losing generality. By introducing these relationships into N-S equations[8] (2.4) and (2.5) and taking the time-average of each term involved in the resulted N-S, we arrive at the Reynolds Averaged Navier-Stokes

---

[7] Other dimensionless numbers: Grash of number for buoyant flow and natural convection, Froude number for free surface flow, and Richardson number for stratified flow.

[8] There is criticism of why a statistical method, like Reynolds decomposition, is associated with deterministic equations, like N-S, instead of stochastic ones [19].

(RANS) equations (2.6) and (2.7) (see Appendix for detailed derivation of RANS) as:

$$\rho\left(\frac{\partial \bar{u}_i}{\partial t} + \bar{u}_j \bar{u}_{i,j}\right) = -\bar{p}_{,i} + \mu \bar{u}_{i,jj} + (R_{ij})_{,j} \qquad (2.6)$$

$$\bar{u}_{i,i} = 0 \qquad (2.7)$$

where $\bar{u}_i$ is the average velocity vector, $\bar{p}$ is average pressure, and
$R_{ij} = -\rho(\overline{u_i' u_j'})$. For a scalar quantity, like temperature $\Theta = \bar{\Theta} + \Theta'$,
the term $-\rho c_p \overline{u_i' \Theta'}$ appears in the averaged heat transfer equa-
tion (see Appendix for detail and definitions). RANS equations are
similar to the original N-S equations (2.4) and (2.5), when the latter
equations are written for averaged quantities, except for the new
additional term on the right-hand side which involves R_{ij}, and makes
these equations indeterminate or unsolvable unless R_{ij} is given in
terms of average velocities. In other words, we need to have a turbulent 'constitutive' relation. This term, i.e., R_{ij} is simply the contribution of the turbulence fluctuations to the non-linear acceleration
terms. It is taken to the right hand side since its dimension is similar
to that of a stress quantity.

We usually, assume that the fluid density fluctuation is small
or $\rho \cong \bar{\rho}$. This assumption is valid for most flows except for buoyant turbulent flows. In general, for flow with about 5% turbulence
intensity and up to Mach number of 5, the fluid density fluctuations
have a small effect on the mean velocity [44], whereas for higher
turbulence intensity, on the order of 20%, for flow with Mach number around 1 the density fluctuations start to affect the mean speed
[40]. For industrial compressible fluids Favre-Averaged N-S equations (**FANS**) [45] are usually used for turbulent modeling. Interested readers are referred to [13] and [21] for detail discussions on
this topic.

There are a couple of important points related to the RANS
equations, that we should focus on before continuing our discussion:

a) In taking the time average of quantities involved,
    the time average for a function $\Phi(x,t)$ is defined as

$$\bar{\Phi}(x) = \lim_{T \to \infty} \frac{1}{T} \int_0^T \Phi(x,t)\,dt. \text{ In other words, we could collect}$$

the turbulent velocity signals, for example, in a flow at a given location using a probe, and then calculate the average value of all the collected data over a time span of T. Then for a 'sufficiently' large value of T (compared to the time scale of fluctuations) we can define an average velocity. Obviously, the averaged value does not depend on time. Now if we replace Φ with velocity $u_i = \bar{u}_i + u'_i$, we have

$$\bar{u}_i(x) = \lim_{T \to \infty} \frac{1}{T} \int_0^T (\bar{u}_i(x) + u'_i(x,t)) dt = \bar{\bar{u}}_i(x) + \overline{u'}_i(x,t) \qquad (2.8)$$

$$= \bar{u}_i(x) + \overline{u'}_i(x,t)$$

Therefore we have two properties resulting from Reynolds decomposition and averaging operations, i.e. $\bar{\bar{u}}_i = \bar{u}_i$ and $\overline{u'}_i = 0$. We should also note that the first term on the left hand side of equation (2.6) (i.e. $\frac{\partial \bar{u}_i}{\partial t}$) should be zero since average velocity is independent of time. However, this term is kept in and there are several justifications [20] for why we should/could have this term included.

b) As mentioned, the new quantity (i.e. $R_{ij} = -\rho \overline{u'_i u'_j}$), which is the result of averaging and decomposition operations, should be calculated or known in order to have a determinate system of equations. Since we have ten unknowns (three components of averaged velocity $(\bar{u}_1, \bar{u}_2, \bar{u}_3)$, pressure, and six components of the symmetric tensor R_{ij}) and four equations (i.e., one for continuity and three for momentum), hence we end up with an indeterminate system. In other words, RANS equations (2.6) and (2.7) need a 'closure' or a relationship between Reynolds stresses and mean velocities.

The quantity $R_{ij} = -\rho \overline{(u'_i u'_j)}$ is now explained and expanded. This quantity has the same dimension as a stress tensor, therefore it is called *Reynolds stress.* When written as $R_{ij} = -\overline{(\rho u'_i) u'_j} = -\rho \overline{(u_i - \bar{u}_i) u'_j} = -\underbrace{\overline{(\rho u_i) u'_j}}_{} + \rho \underbrace{\overline{\bar{u}_i u'_j}}_{=0}$, it shows more

clearly that Reynolds stress is the transport of mean i-component of momentum fluctuation per unit volume, ρu_i, by j-component of velocity fluctuation, u'_j (or vice versa). Obviously, when we have momentum transfer (by Newton's second law) we have force/stress involved, which again indicates that this quantity is actually extra stresses due to turbulence. R_{ij} could also be interpreted as averaged momentum transfer per unit area per unit time due to turbulence fluctuations. Obviously, Reynolds stress depends on the *flow* not the *fluid* properties, which makes it much more challenging to be defined in terms of averaged quantities as compared to laminar flow where the stress could be related to velocity through fluid properties (and in the case of Newtonian fluid through fluid's viscosity).

For a three-dimensional turbulent flow, components of Reynolds stress are:

$$R_{ij} = -\rho \begin{bmatrix} \overline{u_1'^2} & \overline{u_1'u_2'} & \overline{u_1'u_3'} \\ \overline{u_1'u_2'} & \overline{u_2'^2} & \overline{u_2'u_3'} \\ \overline{u_1'u_3'} & \overline{u_2'u_3'} & \overline{u_3'^2} \end{bmatrix} \tag{2.9}$$

As mentioned, this is a symmetric matrix[9] and it has six independent components or unknowns. The diagonal terms are averaged co-correlation of the turbulent velocity fluctuations which are extra (in addition to average pressure) normal 'stresses' or pressures exerted on the fluid due to turbulence. The off-diagonal components are averaged cross-correlation of the turbulent velocity fluctuations which are 'extra' shear stresses (in addition to averaged shear stresses) exerted on the fluid due to turbulence. Another important property of R_{ij} is the trace (i.e. sum of the diagonal terms) of this tensor:

$$R_{ii} = R_{11} + R_{22} + R_{33} = -\rho\left(\overline{u_1'^2} + \overline{u_2'^2} + \overline{u_3'^2}\right) \tag{2.10}$$

The quantity in the bracket, sum of averaged squares of velocity fluctuations, is actually twice the amount of turbulent kinetic energy per unit mass of the fluid, k

$$k = \frac{1}{2}\left(\overline{u_1'^2} + \overline{u_2'^2} + \overline{u_3'^2}\right) = \frac{-R_{ii}}{2\rho} \tag{2.11}$$

Note that k is a positive quantity, as it should be, and hence R_{ij} is negative, as defined here.

[9.] The Reynolds stress is a second order tensor with real eigenvalues. It is a positive semi-definite tensor.

Readers should pay attention to this realization that by making an assumption that turbulence is a 'random' phenomenon (i.e. Reynolds decomposition) and by performing a statistical operation (i.e. averaging process on N-S) we pay the 'price' with the emergence of a new quantity (i.e. Reynolds stress tensor, R_{ij}) and the need to find a relationship for it in terms of the averaged quantities (for example velocity). This 'simple fact' or realization is the seed and foundation for continued advances in turbulence modeling [46].

CLOSURE METHODS

Boussinesq's hypothesis: Boussinesq in 1877 [1] proposed that, at the small length scale, turbulent stress is linearly proportional to the mean strain rate. In other words, we can replace components of R_{ij} with relevant average strain rate including a 'constant' of proportionality. The proportionality constant μ_T is called turbulent *eddy viscosity*. Mathematically, Boussinesq's hypothesis can be written as:

$$R_{ij} = \mu_T(\overline{u}_{i,j} + \overline{u}_{j,i}) - \frac{2}{3}\rho k\delta_{ij} \qquad (2.12)$$

In principle μ_T is a tensor of fourth order (with 81 components which could be reduced to 25 components, due to symmetry [47]), since it relates two second-order tensors to each other. Actually Boussinesq's hypothesis is a constitutive equation for turbulence and, as mentioned previously, could be used to 'close' the RANS equations. The resemblance of Boussinesq's relation (Equation 2.12) and the constitutive relation for a laminar flow (or Newton's law of viscosity, Equation 2.1) is quite interesting! However, for the laminar flow we could use one of fluid's properties, like viscosity, for all flow conditions, whereas treating the turbulence just by introducing a single turbulent eddy viscosity (which is a property of the flow) creates a huge constraint on the resulting turbulence model. In addition, we require calculating the eddy viscosity using other quantities involved in a turbulent flow. To address this constraint or shortcoming, several methods have been developed to calculate eddy viscosity using more information related to turbulence, for example turbulence kinetic energy, using algebraic (including the mixing-length theory of Prandtl [48]) and differential equations. In the next section, we will discuss and explain some of these methods which are commonly used in engineering practice.

Using Boussinesq's hypothesis and the corresponding equation (2.12) for R_{ij}, after some manipulations on Equations (2.6) and (2.7) (see Appendix for more details), we arrive at Equations (2.13) and (2.14) (for incompressible Newtonian fluids):

$$\rho\left(\frac{\partial \overline{u}_i}{\partial t} + \overline{u}_j \frac{\partial \overline{u}_i}{\partial x_j}\right) = -\frac{\partial \overline{p}^\circ}{\partial x_i} + (\mu + \mu_T)\frac{\partial^2 \overline{u}_i}{\partial x_j \partial x_j} \qquad (2.13)$$

$$\overline{u}_{i,i} = 0 \qquad (2.14)$$

where $\overline{p}^\circ = \overline{p} + \frac{2}{3}\rho k$. Obviously, this system of equations is determinate (since we have four unknowns and four equations) once turbulent eddy viscosity μ_T is given. Note that k does not need to be calculated explicitly and is absorbed into the average modified pressure term to give the total pressure $\overline{p}^*$, i.e. average thermodynamic pressure plus extra pressure due to turbulent fluctuations.

Calculating turbulent eddy viscosity, μ_T is the main objective of RANS models which are mostly used in engineering applications. Finally, and before proceeding to discuss these models, we should mention that Boussinesq's hypothesis has some shortcomings [20], [49] (about which he even himself gave some warnings!) from the physics of turbulence point of view. Among these, we may mention the turbulence anisotropy, which exists when normal Reynolds stresses depend on rotational direction, for example, secondary flows in a noncircular duct. The fundamental assumption of Boussinesq's hypothesis implies isotropic turbulent stresses (see Equation 2.12).

Alternative methods: Several other models for calculating Reynolds stresses and hence 'closing' the RANS equations have been developed, and are continuing to be the subject of research. Some of these methods are extensions to the turbulent eddy viscosity model using relatively simpler or more complex equations, including nonlinear equations. On the other hand, some models ignore Boussinesq's hypothesis and make attempts to calculate the Reynolds stresses by using N-S equations directly, for example the Reynolds Stress Model (RSM) [40]. When considering the task of modeling, which is, in general, replacing a complicated system of equations, which govern a phenomenon, with a simpler one that can be solved either analytically or numerically we find that some existing turbulence models do not provide a 'simpler' problem or equations to be solved instead of original equations, especially, for those models which try to calculate the Reynolds stresses directly from N-S equations. This complication results from the fundamental assumptions related to the RANS method, which replaces *physics* with *statistics* [19]. As previously mentioned, other models like LES and DNS are receiving more attention, among researchers, for turbulence modeling in order to overcome these deficiencies. However, RANS models constitute the core of engineering models available for industrial-scale applications in the context of commercial CFD software, like COMSOL, and satisfy, at least in many cases, the need for finding an 'engineering approximation' for industrial turbulent flows [40]. In the author's view, RANS models will remain in application even when more advanced methods, like LES or DNS, become more commercially available and economically viable. The reason could be claimed to be the relatively low computational cost as well as the fact that RANS models may be useful for generating initial values as inputs for more advanced models of complex turbulent flows. One may also predict and anticipate for a major breakthrough in turbulence modeling, for example using DNS approach, through application of quantum computational technologies, during the twenty-first century.

SPECIFIC RANS MODELS

There are several ways to categorize RANS-based models. However, the most common way is to list them according to the type and number of extra equations (in addition to the conservation equations) used for calculating the Reynolds stress, in order to 'close' the RANS system of equations. In a recent publication Alfonsi [50] has reported a comprehensive list of RANS models with their corresponding descriptions. To help with our discussion in this book, we list and briefly explain those models which are available in the COMSOL (versions 4.4 and 5) CFD module.

1. ***First-order models*** are those models which use algebraic equation(s), ordinary differential equation(s)/ODEs, or partial

differential equation(s)/PDEs for calculating Reynolds stress using turbulent eddy viscosity approximation or Boussinesq's hypothesis. There are two fundamental assumptions (as a consequence of Boussinesq's approximation) related to first-order models: (1) the requirement for proper velocity and length scales (or equivalently a time scale), since

$$\nu_T = \frac{\mu_T}{\rho} \equiv \vartheta \ell, \text{ where } \vartheta \text{ is velocity scale, and } \ell \text{ is length scale of}$$

the turbulent flow, and (2) local equilibrium (where turbulent production and dissipation are in balance) and isotropic turbulent flows or the requirement that turbulent eddy viscosity is the same for all directions at any point in the flow. In other words, in an isotropic turbulent flow the ratio of Reynolds stress and the mean strain rate is independent of the rotation of the coordinate system. The latter is not valid if turbulent flow is anisotropic, which is the case for turbulence associated with large eddies. We should also mention that the principle axis of the mean strain rate and that of Reynolds stress are the same when Boussinesq's relation for eddy viscosity is assumed. This is not always the case in some turbulent flows.

The following turbulence models are usually considered under this category:

- **Algebraic (so-called zero equation) models:** For these models we use algebraic relation(s) (but no PDE, hence the name zero-equation model) to calculate the eddy viscosity, and hence Reynolds stress components, to close the RANS equations. Therefore, only a system of four PDEs for the average velocity and pressure is solved. Examples are the constant eddy viscosity model, Prandtl's model [48] which is based on his mixing-length theory[10] (more reliable for 2D free shear flows and flows with mild slope without boundary layer separation), the Cebeci-Smith model [51] (applicable to wall-bounded 2D flows without separation), and the Baldwin-Lomax model [52] (applicable to 3D flows without separation). In summary, the zero-equation

---

[10] Some references [13] distinguish between mixing-length and eddy-viscosity/Boussinesq approaches, although these can be 'mixed' to model the turbulent viscosity, in general.

models provide good engineering solutions for the types of turbulent flows that are calibrated for, as mentioned above, and when relevant and correct modeling constants are used. Two Algebraic models are available in COMSOL version 5 (not available in version 4.4). These models are the Algebraic yPlus model (based on Prandtl's mixing-length theory) and the L-VEL model of Agonafer [53] (based on an extension of logarithm law of the wall).

- **One-equation models:** For these models we use a PDE which governs transport of, for example, turbulent kinetic energy k or eddy viscosity. The dependence of eddy viscosity on the kinetic energy and its transport through turbulent flow brings more accountability to the model since some non-local effects and turbulent flow history is accounted for with the use of a proper velocity scale (i.e. $\sqrt{k}$). In addition, we should use a proper length scale for reliable modeling. In this group a system of five PDEs for the average velocities, pressure, and kinetic energy is solved. Examples are the Baldwin-Barth model, and the Spalart-Allmaras (S-A) model, which uses a transport equation for eddy viscosity. Among these models S-A can predict more reliable results for separated flows related to bounded aerodynamics and turbo-machinery applications. The Spalart-Allmaras model is available in COMSOL (versions 4.4 and 5).

- **Two-equation models:** For these models we have two PDEs which govern turbulence quantities, for example kinetic energy k and energy dissipation rate ε, and are used to calculate eddy viscosity. The dependence of eddy viscosity on transport of turbulence quantities brings more accountability to the model since more proper turbulent velocity (i.e $\sqrt{k}$ and length i.e.$\sqrt{k^3}/\varepsilon$) scales are involved. A system of six PDEs for the average velocities, pressure, and two turbulence quantities is solved. Examples are the k-ε model (including its versions), the Low-Reynolds k-ε model, the k-ω model (including its versions), the Shear Stress Transport (SST) model, the non-linear eddy-viscosity model. This group of turbulence models is more popular and mostly used for engineering applications. Two-equation models are relatively more robust with well-developed 'universal' constants associated with each model, a characteristic which is useful in practical applications. Among these models the *k-ε*, *k-ω*, SST, and Low-Reynolds *k-ε* models are available in COMSOL (versions 4.4 and 5).

2. *Second-order models* are those models which use additional PDEs to calculate the Reynolds stresses, mostly directly, from N-S equations, without using eddy viscosity (i.e. Boussinesq's approximation). These types of models, however, overcome some shortcomings resulting from Boussinesq's hypothesis and mixing-length theory applications, and take care of anisotropy in turbulence (e.g. strong curvature, swirling flows, strong buoyancy) but are more costly in terms of computer time, and much more complex for implementing them to engineering applications. Examples of turbulence models considered under this category are:

• **Reynolds stress model:** This model is the most complex one among all models, when the Reynolds averaging method is used. The objective is to derive equations for transport of six independent components of Reynolds stress and an equation for turbulent energy dissipation. The result is a total of seven equations, in addition to conservation equations, which should be solved to model turbulence. Boussinesq's hypothesis is not used in the Reynolds stress model, however, similar approximations are used for modeling some of the terms involved in governing equations [20]. Overall this model is much more demanding of computer time as compared to first-order models and not 'appealing' for engineering applications. This model is not available in the COMSOL CFD module.

• **Algebraic stress model:** This model is an attempt to keep the merits of the Reynolds stress model (e.g. effects of anisotropy in turbulence) but to reduce the computer time and modeling efforts required. The resulting equations are a set of algebraic equations which should be solved along with conservation equations. The fundamental point about these models is based on the assumption that Reynolds stresses are series expansions of functions which include several terms in the expansion. This treatment is similar to what is used in Boussinesq's eddy viscosity approximation, where only the first term in the expansion is considered. Several modeling methods, mainly for gradient of Reynolds stresses, are proposed by researchers for Algebraic stress models [54], [20]. This model is not available in the COMSOL CFD module.

The development and success, of all turbulence models (for example those mentioned in this section) depends on how much they can capture or approximate the wide range of length and velocity scales which exist in turbulent flows. Table 2.1 summarizes some of the commonly used turbulence models, including those discussed above and those available in COMSOL.

TABLE 2.1: Specific RANS models and categories, including those available in COMSOL.

First-order model	Algebraic or Zero-equation	Cebeci-Smith (1967)	N/A
		Baldwin-Lomax (1978)	N/A
		Algebraic yPlus	Available in COMSOL, V5.
		L-VEL (1996)	Available in COMSOL, V5.
	One-equation	Baldwin-Barth (1990)	N/A
		Spalart-Allmaras (1992)	Available in COMSOL, V4.4&5
	Two-equation	k-ε (1974) and Low-Re version	Available in COMSOL, V4.4&5
		k-ω (versions 1972–2006)	Available in COMSOL, V4.4&5
		SST (1993, and its variations)	Available in COMSOL, V4.4&5
Second-order model	Reynolds stress model (RSM)		N/A
	Algebraic stress model (ASM)		N/A

COMSOL RANS MODELS

Overview and Governing Equations

In this section, we cover the governing equations, discussions on boundary conditions, and merits and shortcomings related to seven RANS models which are currently available in the COMSOL CFD module. These models are as follows:

1. k-ε (k-epsilon) model

2. k-ω (k-omega) model

3. SST (Shear Stress Transport) model

4. Low-Re k-ε (low Reynolds number k-epsilon) model

5. Spalart-Allmaras model

6. Algebraic yPlus model

7. L-VEL model

All of these models belong to the first-order models category, as described in the previous section. Among these the k-ε, k-ω, SST, and Low-Re k-ε models belong to the two-equation models group and make use of Boussinesq's hypothesis for modeling Reynolds stresses which require calculation of turbulent kinematic eddy viscosity $v_T = \dfrac{\mu_T}{\rho}$. Therefore a velocity scale ϑ and a length scale ℓ should be defined and calculated for each model, accordingly (since $v_T \sim \vartheta\ell$). The velocity scale is calculated using $\sqrt{k}$, or the square root of turbulent kinetic energy and length scale is calculated either by using turbulent energy dissipation rate ε (i.e. $\ell \sim \sqrt{k^3}/\varepsilon$) or inverse of the time scale for turbulent large eddies ω (i.e. $\ell \sim \sqrt{k}/\omega$). The resulting eddy viscosity would be $v_T \sim k^2/\varepsilon$ for the k-ε model and $v_T \sim k/\omega$ for the k-ω model.

The Spalart-Allmaras model belongs to the one-equation models group and makes use of a single PDE for calculating kinematic eddy viscosity and employs Boussinesq's hypothesis for modeling Reynolds stresses. The length scale for this model is usually the distance to the closest wall boundary.

The yPlus and L-VEL models belong to the zero-equation models group, where the turbulent viscosity is modeled using algebraic equations. These models are much less demanding on computer power and provide fast results, in terms of computational times required, relative to the other models mentioned above. The yPlus (mostly suitable for internal flows) is based on Prandtl's mixing-length theory and the L-VEL model, also referred to as automatic algebraic method, is an extension to the logarithm law of the wall. The name of the latter is taken from the requirement of the distance to the nearest wall (L) and the local velocity (VEL) for this model, (hence L-VEL!). The L-VEL model gives a good prediction of the turbulent viscosities near the wall (but not as good in regions far from the wall in the main stream) and is suitable for turbulence modeling of problems that includes walls with a wide range of length scales, for example, an electronic equipment cabinet, and urban-area air boundary layer. The yPlus and L-VEL models are available in COMSOL, version 5.

In the next section, we discuss the exact equation for turbulent kinetic energy (which is used for calculating proper velocity scale) and modeling approaches for some of the terms involved in this equation that are suggested to 'close' the system of equations. This is an essential step for understanding the assumptions made for these turbulence models, and hence the merits of two-equation models included in the above-mentioned list. For detail derivation of governing equations readers are referred to the Appendix.

Exact Turbulent Kinetic Energy Transport Equation

In the previous section, we defined the symmetric Reynolds stress, R_{ij}. For our discussion here, we repeat the related equation here again:

$$R_{ij} = -\rho \begin{bmatrix} \overline{u_1'^2} & \overline{u_1'u_2'} & \overline{u_1'u_3'} \\ \dots & \overline{u_2'^2} & \overline{u_2'u_3'} \\ \dots & \dots & \overline{u_3'^2} \end{bmatrix} \tag{2.15}$$

where (...) indicates the symmetry for corresponding terms. The trace of Reynolds stress is $R_{ii} = -\rho\left(\overline{u_1'^2} + \overline{u_2'^2} + \overline{u_3'^2}\right)$. Turbulent kinetic energy per unit mass (or specific kinetic energy)[11] is defined as $k = \frac{1}{2}\left(\overline{u_1'^2} + \overline{u_2'^2} + \overline{u_3'^2}\right)$, a non-negative quantity. These relationships clearly show that in a turbulent flow kinetic energy is related to the normal components of Reynolds stress $\left(\text{i.e.} = \frac{-R_{ii}}{2\rho}\right)$. One should recall that the origin of Reynolds stress was the application of Reynolds decomposition and time-averaging operations on N-S equations. Therefore, inclusion of turbulent kinetic energy in calculation of Reynolds stresses, which are approximated through Boussinesq's hypothesis and eddy viscosity, is of great importance to turbulent model fidelity. As mentioned in the previous section, we use k to estimate the turbulence velocity scale. The exact equation for transport of k can be derived in a number of ways, for example by performing a contraction (i.e. $\delta_{ij} R_{ij} = R_{ii}$) operation on the transport equation for R_{ij} [13] or using the governing equation for velocity perturbations [21]. The related derivations are tedious and lengthy (see the Appendix) but here we discuss the physical significance of relevant terms involved in the governing equations for R_{ij} first, which are quite important for readers' understandings of these terms and in order to properly use turbulent models.

The transport equation for Reynolds stress R_{ij} is given as follows (see Appendix):

$$\frac{\partial R_{ij}}{\partial t} + \overline{u}_k \frac{\partial R_{ij}}{\partial x_k} = P_{ij} + \phi_{ij} - \psi_{ijk} + D_{ij} - \varepsilon_{ij} \tag{2.16}$$

The terms on the left-hand side of equation (2.16) represent the time rate change and convection of the Reynolds stress tensor. The terms on the right hand side are:

1. The Reynolds stress production rate by mean velocity gradient $P_{ij} = R_{ik} \frac{\partial \overline{u}_j}{\partial x_k} + R_{jk} \frac{\partial \overline{u}_i}{\partial x_k}$. This term does not require further modeling and is used as is for the calculation of Reynolds stresses.

---

[11] Unless otherwise mentioned, all references to quantity k is meant to be turbulent-specific kinetic energy.

2. The correlation between the fluctuating pressure and the fluctuating strain rate $\phi_{ij} = \overline{p'\left(\dfrac{\partial u'_i}{\partial x_j} + \dfrac{\partial u'_j}{\partial x_i}\right)}$. It is obvious that, for incompressible fluids, the trace of this term (or ϕ_{ii}) is zero and hence it does not contribute directly to the kinetic energy. Its contribution is important for stresses in turbulence anisotropy. This term requires modeling.

3. The combination of turbulent diffusion correlation, which all contribute to the re-distribution of Reynolds stresses through the fluid domain $\psi_{ijk} = \dfrac{\partial}{\partial x_k}\left(\rho\overline{u'_i u'_j u'_k} + \overline{p'u'_i}\delta_{jk} + \overline{p'u'_j}\delta_{ik}\right)$.
   This term requires modeling.

4. The diffusion of Reynolds stresses or viscous transport $D_{ij} = \nu\dfrac{\partial^2 R_{ij}}{\partial x_k \partial x_k}$. This term does not require modeling.

5. The dissipation rate of Reynolds stresses through fluid viscosity, $\varepsilon_{ij} = 2\nu\overline{\dfrac{\partial u'_i}{\partial x_k}\dfrac{\partial u'_j}{\partial x_k}}$. This term requires modeling.

The Transport equation for Reynolds stresses consist of six independent equations for six components of the symmetric Reynolds stress tensor. Unfortunately, solution of these equations requires modeling of a total of 22 unclosed statistical terms involved [19]. This requirement and the high demand for computational power and time are major disadvantages of the Reynolds Stress modeling approach, categorized as second-order models and mentioned in the previous section. In this book we do not cover the Reynolds stress model nor is this model available in COMSOL. However, the equation for Reynolds stress transport is used for derivation of turbulent kinetic energy. The k equation (2.17) for an incompressible Newtonian fluid can be derived [22] from the transport equation (2.16) (see Appendix):

$$\rho\left(\frac{\partial k}{\partial t} + \overline{u}_j\frac{\partial k}{\partial x_j}\right) = P_k - \mu\overline{\frac{\partial u'_i}{\partial x_j}\frac{\partial u'_i}{\partial x_j}} + \mu\frac{\partial^2 k}{\partial x_j \partial x_j} \qquad (2.17)$$
$$-\frac{\partial}{\partial x_j}\left(\frac{1}{2}\rho\overline{u'_i u'_i u'_j} + \overline{p'u'_j}\right)$$

The terms on the left-hand side represent the time rate change and convection of kinetic energy by mean stream, respectively, i.e. total rate of change of turbulent kinetic energy. The terms on the right hand side are:

1. The production or source term $P_k = R_{ij}\dfrac{\partial \overline{u}_i}{\partial x_j}$, representing the turbulent kinetic energy transferred from the mean flow. In other words, it is turbulent kinetic energy per unit time per unit volume that is gained by an eddy due to mean strain rate. This term requires modeling.

2. The dissipation (or rather isotropic dissipation [13]) term or sink term, $\mu\dfrac{\partial u'_i}{\partial x_j}\dfrac{\partial u'_i}{\partial x_j}$, representing mean conversion rate of turbulent kinetic energy per unit volume of the fluid to thermal energy through viscosity. In other words, it represents transfer of mean rate of energy transferred by the smallest eddies (Kolmogorov scale) to internal energy. This term requires modeling.

The remaining terms are of diffusion types:

3. The diffusion or redistribution term, $\dfrac{\partial^2 k}{\partial x_j \partial x_j}$, representing diffusion of turbulent kinetic energy flux. The divergence from $\left(\text{i.e.}\dfrac{\partial}{\partial x_j}\left(\mu\dfrac{\partial k}{\partial x_j}\right)\right)$ shows, more clearly, that turbulent kinetic energy flux is diffused through a fluid's dynamic viscosity or molecular diffusion. This term does not require modeling.

4. The last quantity, on the right hand side is actually a combination of two terms as follows. Both of these terms require modeling. The first is (a) $\dfrac{\partial}{\partial x_i}\left(\dfrac{1}{2}\rho\overline{u'_i u'_i u'_j}\right)$, representing the rate of redistribution or diffusion of turbulent kinetic energy flux due to velocity fluctuations or transport of turbulence energy by turbulence, and second is (b) $\dfrac{\partial}{\partial x_j}(\overline{p'u'_j})$, or the pressure diffusion term representing diffusion of turbulent flow work.

Overall, the transport equation for k indicates that if we could 'ride' on a differential control volume of the fluid in a turbulent flow (i.e. a Lagrangian point of view), then the changes that we observe for its total turbulent specific kinetic energy, which are represented by the terms on the left-hand side of equation (2.17), should be in balance with its turbulent kinetic energy gain (from the mean flow), its turbulent kinetic energy destruction (through dissipation), and redistribution (through the fluid's viscosity and velocity-pressure fluctuations).

As mentioned above, in order to derive the equation for k which could be employed for computations, we should model some of the terms appearing in the k-equation (2.17). These terms are production, dissipation, and diffusion due to turbulent fluctuations, i.e. items (1), (2), and (4) as mentioned in the previous list.

For modeling the production term, P_k we employ Boussinesq's hypothesis for Reynolds stresses or equation (2.12). For modeling the dissipation term, usually it is assumed that the local isotropic property for turbulence (at small-scale eddies) exists and hence this term is modeled by $\mu \overline{\dfrac{\partial u'_i}{\partial x_j} \dfrac{\partial u'_i}{\partial x_j}} = \rho\varepsilon$, where ε is the turbulent kinetic energy dissipation rate per unit mass. The local isotropy assumption has major implications for turbulence modeling and is well studied by Wilcox [20]. And finally, the last term is modeled as the gradient of a diffusion term with introduction of a new closure constant parameter, σ_k. Therefore we have $\dfrac{\partial}{\partial x_j}\left(-\dfrac{1}{2}\rho\overline{u'_i u'_i u'_j} - \overline{p' u'_j}\right) = \dfrac{\partial}{\partial x_j}\left(\dfrac{\mu_T}{\sigma_k}\dfrac{\partial k}{\partial x_j}\right)$.

When collecting all of the modelled terms, after rearranging, we get the CFD form of the k-Equation (2.18), for Newtonian incompressible fluids, as

$$\rho\left(\frac{\partial k}{\partial t} + \overline{u}_j \frac{\partial k}{\partial x_j}\right) = \frac{\partial}{\partial x_j}\left(\left(\mu + \frac{\mu_T}{\sigma_k}\right)\frac{\partial k}{\partial x_j}\right) + \overline{P}_k - \rho\varepsilon \qquad (2.18)$$

where $\overline{P}_k = \mu_T \dfrac{\partial \overline{u}_i}{\partial x_j}\left(\overline{u}_{i,j} + \overline{u}_{j,i}\right), \varepsilon \sim \sqrt{k^3}/\ell,$ and $\mu_T \sim \rho\ell\sqrt{k}.$

Note that the term, $\dfrac{2}{3}\rho k\delta_{ij}\dfrac{\partial \overline{u}_i}{\partial x_j} \sim \dfrac{\partial \overline{u}_i}{\partial x_i} = 0$ for incompressible fluids

and $\overline{P}_k$ represents the 'incompressible' part of P_k (see Equation 2.17).

When one-equation models are used, equation (2.18) is solved for k and the turbulence length scale ℓ should be defined, as well. However, for two-equation models, like the k-ε model, another PDE for energy dissipation ε is used for defining the turbulent length scale, along with some constraints that will be discussed later in further sections. In the next section, we discuss the governing equation for ε.

Exact Turbulent Kinetic Energy Dissipation Transport Equation

The turbulent energy dissipation rate per unit mass, ε, appears in equation (2.18), which governs k. Therefore, we need to have the governing equation for transport of ε. This is given by equation (2.19) and can be derived (see [13], [21]) after some tedious and lengthy manipulations using N-S equations (see Appendix):

$$\rho\left(\frac{\partial \varepsilon}{\partial t} + \overline{u}_j \frac{\partial \varepsilon}{\partial x_j}\right) = P_\varepsilon + D_\varepsilon - \Phi_\varepsilon + \frac{\partial}{\partial x_j}\left(\mu \frac{\partial \varepsilon}{\partial x_j}\right) \qquad (2.19)$$

Terms on the left-hand side represent time rate change and convection of ε, respectively. The terms on the right-hand side are [50]:

1. The production or source term P_ε, represents the production of dissipation which consists of dissipation by averaged strains, inhomogeneity, and vortex stretching. This term requires modeling. The exact equation for
$$P_\varepsilon = -2\mu\left[\overline{(u'_{i,k}u'_{j,k})}\overline{u}_{i,j} + \overline{(u'_{i,k}u'_{i,j})}\overline{u}_{j,k} + \overline{(u'_j u'_{i,k})}\overline{u}_{i,jk} + \overline{(u'_{i,k}u'_{j,k}u'_{i,j})}\right]$$

2. The diffusion term, D_ε, represents diffusion of turbulent kinetic energy dissipation which consists of pressure-velocity fluctuations and turbulent velocity fluctuations. This term requires modeling. The exact equation for
$$D_\varepsilon = -2\nu\left(\overline{u'_{i,k}p'_k}\right)_{,i} - \mu\left(\overline{u'_j u'_{i,k}u'_{i,k}}\right)_{,j}$$

3. The dissipation (or destruction) term Φ_ε, represents the dissipation of turbulent kinetic energy dissipation. This term requires modeling. The exact form for $\Phi_\varepsilon = 2\mu\nu\left(\overline{u'_{i,jk}u'_{i,jk}}\right)$.

4. The viscous transport term $\dfrac{\partial}{\partial x_j}\left(\mu\dfrac{\partial \varepsilon}{\partial x_j}\right)$, represents the

diffusion or redistribution of ε through fluid dynamic viscosity. This term does not require modeling.

Overall, the transport equation for ε indicates that if we could 'ride' on a differential control volume of the fluid in a turbulent flow (i.e. a Lagrangian point of view), then the changes that we observe for its total turbulent specific kinetic energy dissipation, which are represented by the terms on the left-hand side of equation (2.19), should be in balance with its turbulent kinetic energy dissipation production (to the mean flow) plus its turbulent kinetic energy dissipation destruction (through dissipation) and redistribution (through the fluid's viscosity, velocity, and pressure fluctuations).

As mentioned above, in order to derive the equation for ε, which can be employed for computations, we should model (see [13], [50]) some of the terms appearing in the ε equation (2.19). These terms are production, dissipation, and diffusion due to turbulent fluctuations, i.e. items (1), (2), and (3) as mentioned in the list above. For modeling the production term, we assume local equilibrium (i.e. turbulent kinetic energy and its dissipation are balanced) which implies that production of ε is governed by Reynolds stress and average velocity gradient. This results in $P_\varepsilon = C_{\varepsilon 1}\dfrac{\varepsilon}{k}R_{ij}\dfrac{\partial \overline{u}_j}{\partial x_i} = C_{\varepsilon 1}\dfrac{\varepsilon}{k}\overline{P}_k$. $C_{\varepsilon 1}$ is a non-dimensional constant. The diffusion term is modeled using a transport hypothesis using turbulent eddy viscosity, which results in $D_\varepsilon = \dfrac{\partial}{\partial x_j}\left(\dfrac{\mu_T}{\sigma_\varepsilon}\dfrac{\partial \varepsilon}{\partial x_j}\right)$, where σ_ε is a non-dimensional constant. The dissipation or destruction term is modelled by realizing that this term represents the time rate $\left(\tau = \dfrac{k}{\varepsilon}\right)$ at which ε is destroyed. Therefore we have $\Phi_\varepsilon \sim \dfrac{\varepsilon}{\tau} = C_{\varepsilon 2}\rho\dfrac{\varepsilon^2}{k}$, where $C_{\varepsilon 2}$ is a non-dimensional constant. By substituting these modelled terms into equation (2.19), after rearrangement, we have the CFD form of ε, equation (2.20) (for incompressible fluids) as:

$$\rho\left(\frac{\partial \varepsilon}{\partial t} + \overline{u}_j\frac{\partial \varepsilon}{\partial x_j}\right) = \frac{\partial}{\partial x_j}\left(\left(\frac{\mu_T}{\sigma_\varepsilon} + \mu\right)\frac{\partial \varepsilon}{\partial x_j}\right) + C_{\varepsilon 1}\frac{\varepsilon}{k}\overline{P}_k - C_{\varepsilon 2}\rho\frac{\varepsilon^2}{k} \quad (2.20)$$

Now we have all the equations, in CFD format, required for the first COMSOL RANS model or the k-ε model. In the next section we discuss the details of this model and cover the related equations for Newtonian incompressible fluids. The COMSOL interface allows modeling for compressible fluids (Mach number Ma < 0.3 and high Mach number flow [55]). A detailed discussion for compressible fluids can be found in [21].

k-ε MODEL

In this section, we summarize all the equations for the k-ε model and compare them with those available in the COMSOL CFD module [55]. We also discuss general aspects of boundary conditions for k and ε, related constraints, and types of turbulent flows for which the k-ε model could be applied with acceptable engineering 'accuracy'. In Chapter 4, we apply the COMSOL k-ε model to some problem examples.

The standard k-ε model was developed in the 1970s and is one of the most used, analyzed, and validated turbulence models in industry and academia. Therefore its strengths and weaknesses are well documented [40]. Further modifications came through development of several versions of the standard k-ε model, for example the two-layer k-ε model [56], and the RNG (renormalization group) k-ε model [57], the latter of which is based on a statistical mechanics approach to model the eddy viscosity. For the k-ε model in total, we have six coupled PDEs which govern six variables, i.e. four equations for average velocity vector components and pressure and two equations for k and ε. In order to solve the system of PDEs we need to have the numerical values of the parameters which are defined for modeling different terms involved in these equations. The k-ε model equations (2.21–24), for incompressible Newtonian fluids, are those for continuity, momentum, kinetic energy, and kinetic energy dissipation (which are repeated here for convenience):

$$\overline{u}_{i,i} = 0 \tag{2.21}$$

$$\rho\left(\frac{\partial \overline{u}_i}{\partial t} + \overline{u}_j \frac{\partial \overline{u}_i}{\partial x_j}\right) = -\frac{\partial \overline{p}^*}{\partial x_i} + (\mu + \mu_T)\frac{\partial^2 \overline{u}_i}{\partial x_j \partial x_j} \tag{2.22}$$

$$\rho\left(\frac{\partial k}{\partial t} + \overline{u}_j \frac{\partial k}{\partial x_j}\right) = \frac{\partial}{\partial x_j}\left(\left(\mu + \frac{\mu_T}{\sigma_k}\right)\frac{\partial k}{\partial x_j}\right) + \overline{P}_k - \rho\varepsilon \qquad (2.23)$$

$$\rho\left(\frac{\partial \varepsilon}{\partial t} + \overline{u}_j \frac{\partial \varepsilon}{\partial x_j}\right) = \frac{\partial}{\partial x_j}\left(\left(\mu + \frac{\mu_T}{\sigma_\varepsilon}\right)\frac{\partial \varepsilon}{\partial x_j}\right) + C_{\varepsilon 1}\frac{\varepsilon}{k}\overline{P}_k - C_{\varepsilon 2}\rho\frac{\varepsilon^2}{k} \quad (2.24)$$

where $\overline{p}^* = \overline{p} + \frac{2}{3}\rho k$, $\overline{P}_k = \mu_T \frac{\partial \overline{u}_i}{\partial x_j}\left(\overline{u}_{i,j} + \overline{u}_{j,i}\right)$, $\mu_T = \rho C_\mu \frac{k^2}{\varepsilon}$, $C_\mu =$

0.09, $C_{\varepsilon 1} = 1.44$, $C_{\varepsilon 2} = 1.92$, $\sigma_k = 1.0$, and $\sigma_\varepsilon = 1.3$

The model is sensitive to the numerical value of C_μ, which defines the level of turbulence. It is also sensitive to the difference $(C_{\varepsilon 1} - C_{\varepsilon 2})$ value, which determines the production and dissipation of ε (as can be seen from ε-equation 2.24). The model parameters are optimized and should not be altered, unless validated or experimental data are available in support. The corresponding k-ε model equations (in vector notation) used in COMSOL are shown in Figure 2.3. For comparison, readers can identify equivalent terms in equations and benefit from the discussions given in the previous sections and this section, for their physical interpretations and corresponding assumptions when using this model. For converting tensor notation to vector notation (or vice versa) see [41].

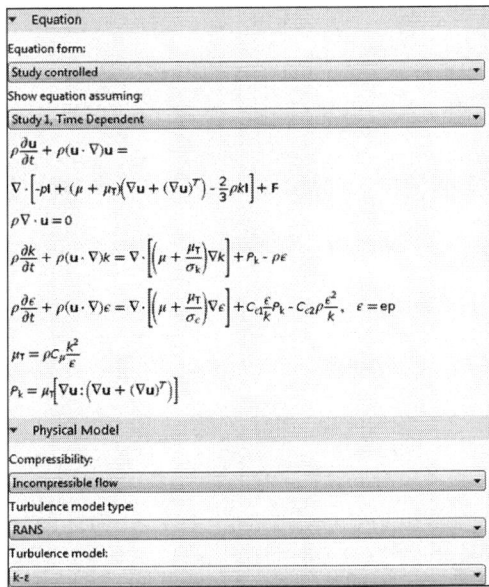

FIGURE 2.3: COMSOL $_k\varepsilon$ model equations for incompressible fluids.

We now discuss several useful and important points related to analysis, properties, and application of the k-ε model.

1. Close to and at the solid boundary wall the transport equation for ε creates some challenges. Two terms on the right-hand side of equation (2.24), the production term $\left(\sim \dfrac{\varepsilon}{k}\right)$, and destruction term $\left(\sim \dfrac{\varepsilon^2}{k}\right)$, become very large close to the wall, or mathematically singular. By definition, k vanishes at the solid wall with a no-slip boundary condition assumption but its dissipation ε doesn't. Therefore, close to a solid wall high resolution mesh is needed for numerical computations to capture the detail of the boundary layer and hence at a solid wall this model does not apply, in general. In other words, k-ε model equations cannot be integrated up to a solid wall. In COMSOL, this limitation is referred to as *mixing-length limit*. To resolve this difficulty a wall function is usually used for computations close to the wall. Wall functions are approximations and for some turbulent flows more accurate modeling is required for the boundary layer and capturing flow details close to a wall. For example, for internal flows usually wall functions are adequate but for external flow or when heat transfer is important more accurate modeling methods are used. We will discuss this topic in a further section. An approach for more accurate modeling is to modify the standard k-ε model equations such that they can be integrated/solved up to a solid wall. This approach will lead to the *Low-Re k-ε* model [58], which requires relatively high resolution mesh near solid walls. We will discuss the Low-Re k-ε model in a further section.

2. The requirement of having non-negative normal Reynolds stresses (i.e. $\overline{\rho u_i'^2}$) could be violated when mean strain becomes large enough. Or similarly the shear Reynolds stresses (i.e. $\overline{u_i'u_j'}$) could violate Schwarz inequality (i.e. $\left(\overline{u_\alpha'u_\beta'}\right)^2 \leq \left(\overline{u_\alpha'^2}\right)\left(\overline{u_\beta'^2}\right)$, with no summation over indices α and β, for large enough mean strain. To avoid such modeling results, which do not make sense from the turbulence physics

point of view, a constraint, the so called *realizability* constraint [59], is imposed on the eddy viscosity μ_T. The realizability constraint specifies that the modeled Reynolds stress tensor should have non-negative eigenvalues and satisfy Schwarz inequality. In other words all the diagonal components of the Reynolds stress tensor remain non-negative and the off-diagonal components satisfy Schwarz inequality. The k-ε model available in COMSOL is a *realizable* one. Readers may also refer to [20] for a more detailed explanation of realizability constraint.

3. Model parameters ($C_\mu = 0.09$, $C_{\varepsilon 1} = 1.44$, $C_{\varepsilon 2} = 1.92$, $\sigma_k = 1.0$, and $\sigma_\varepsilon = 1.3$) or coefficients are determined by applying semi-empirical and optimization methods and are 'universal' [54]. This is considered as the strength of the k-ε model which makes this model employable for many different flows using the same coefficients. However, users should always consider the foundation and assumptions based on which the model is built, as discussed in this section, and evaluate their numerical results as much as possible with relevant references.

4. The equations for the k-ε model are non-linear and coupled, k and ε equations (2.23) and (2.24) are highly coupled with each other and relatively lightly with the conservation equation (2.22) for momentum. This requires proper numerical methods in application for integrating these equations, which is resolved in COMSOL by using proper solvers.

5. The k-ε model relies on several assumptions, the most important of which is that the Reynolds number is high enough. It is also important that the turbulence is in equilibrium in boundary layers, which means that turbulence production equals its dissipation. These assumptions limit the accuracy of the model, because they are not always valid. This model does not, for example, accurately model flows with adverse pressure gradients in the boundary layer and can result in under-predicting the spatial extension of recirculation zones [20]. Furthermore, in the cases of rotating flows and flows over curved boundaries, the model

often shows poor agreement with experimental data [60]. However, this model is nevertheless the most widely used and validated turbulence model, with relatively good performance for many industrial flows. In most cases, the limited accuracy is a fair trade-off for the amount of computational resources saved and the good convergence rate compared to more complicated turbulence models.

Boundary Conditions: Assigning boundary conditions for velocity and pressure is relatively straight forward. For values of k and ε at the symmetry plane, free surface, and outlet flow Neumann-type boundary conditions (i.e. $\frac{\partial k}{\partial n} = 0$ and $\frac{\partial \varepsilon}{\partial n} = 0$) can be used. However at inflow and outflow boundaries their values are not usually known. For inflow boundary condition turbulence intensity I_T and length scale L_T could be used. Turbulent intensity is a measure of the ratio of turbulent flow fluctuations to its mean value. For example the turbulence intensity for velocity u is $I_T = u_{rms}/\bar{u}$, where $u_{rms} = \sqrt{\overline{u'^2}}$ or the root-mean-square of velocity fluctuations for a period of time, as shown in Figure 2.4.

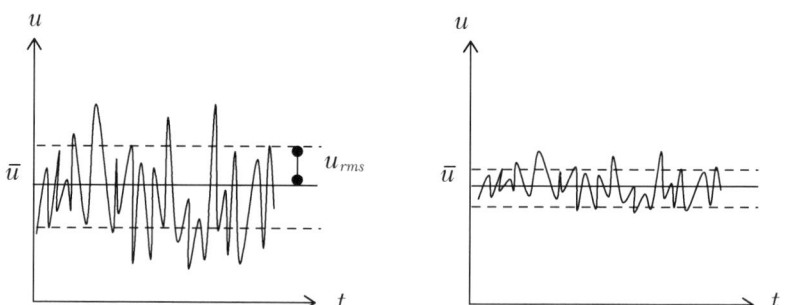

FIGURE 2.4: Turbulence level and 'rms' for two velocity signals with equal means.

Typical values for turbulence intensity for fully developed turbulent flow is about 5–10% and for flows with weak turbulence is about 0.1%. The length scale is selected depending on the type of flow, as shown in Table 2.2. In COMSOL, as shown in Figure 2.5, users also have the option to choose Dirichlet type (i.e. explicit values) boundary conditions for k and ε at the inlet. In the author's view since these are, in principle, statistical values rather than physical ones, using this option for boundary conditions should be performed with care and reliable supporting data.

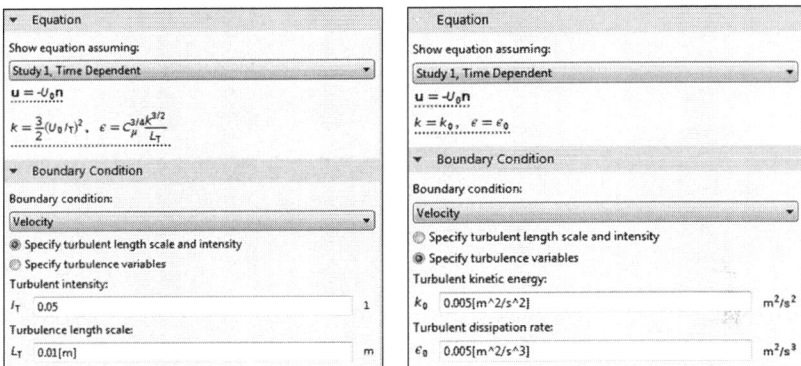

FIGURE 2.5: Typical boundary conditions for turbulent inflow in the COMSOL k-ε model.

For boundary conditions at solid walls, wall functions are used for a region close to the wall or boundary layer, and the k-ε model is assumed to start from a distance δ_W away from the wall as shown schematically in Figure 2.6. In COMSOL, δ_W (the so-called viscous/buffer sub-layer which is small compared to the dimensions of the flow domain geometry) is automatically calculated such that $\delta_W^+ = \rho C_\mu^{0.25} \sqrt{k} \delta_W / \mu$ becomes equal to 11.06. The value of δ_W^+ is available, by default, in the COMSOL modeling results and it is recommended to check this value against 11.06 along the solid walls. If the δ_W^+ value exceeds largely from the 11.06 limit, then a finer mesh should be sued.

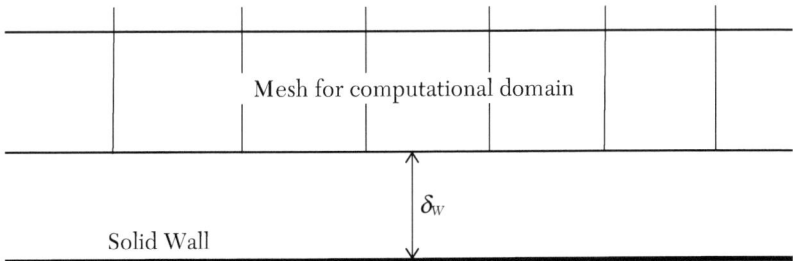

FIGURE 2.6: Sketch for the wall functions domain within a distance δ_W from a solid wall and computational domain for the k-ε model.

TABLE 2.2: Typical values for turbulent length scale for 2D flows (depicted from the COMSOL manual [55]).

Flow Type	L_T	L
Mixing layer	0.07L	Layer width
Plane jet	0.09L	Jet half width
Wake	0.08L	Wake width
Axisymmetric jet	0.075L	Jet half width
Pipe and channels (fully developed flows)	0.07L	Pipe radius or channel half width
Boundary layer	see COMSOL manual	

Initial Values: As discussed in the previous sections, time variation for averaged velocity in a turbulent flow, by definition, should not exist. However, since the mean flow field may vary, like in a transition phase, we may require solving for unsteady RANS equations [20]. In COMSOL default initial conditions for both steady state and transient turbulent flows are set. For details, readers are referred to the COMSOL manual [55].

k-ω MODEL

In this section, we summarize all equations for the k-ω model and compare them with those available in the COMSOL CFD module. We also discuss general aspects of boundary conditions for k and ω as well as the type of turbulent flows for which the k-ω model could be applied with acceptable engineering 'accuracy.' In Chapter 4, we apply the COMSOL k-ω model and discuss details of problem examples.

The k-ω model, was developed about the same time as the k-ε, its development can be traced back to the works of Kolmogorov, Prandtl, Saffman, and Wilcox [61]. This model shows better results compared to those of the k-ε model for separated flows, jet flow, and flows with adverse pressure gradient. Generally speaking, its development was caused by the shortcomings of the k-ε model, for example close to a wall, as well as the complexity of defining the value of ε at the wall. The velocity scale in the k-ω model is $\sqrt{k}$, similar to the k-ε model, but for the length scale ℓ turbulence frequency *ω is used,* i.e. $\ell = \sqrt{k}/\omega$. It would be useful to compare these two models from the point of turbulence quantities employed for defining velocity and length scales. As shown, schematically in Figure 2.7, a typical swirl or eddy carries the turbulent kinetic energy per unit mass, k, and dissipates it per unit time, ε. One could also measure the dissipation rate per unit kinetic energy, i.e. $\omega = \varepsilon/k$. In other words, in the k-ω model the two additional governing equations are the transport equation for k and that for the *rate* at which the kinetic energy dissipates. It is obvious, using dimensional analysis, that the dimension of ω is the inverse of time or it can be interpreted as the inverse of the time scale of large-scale turbulent eddies.

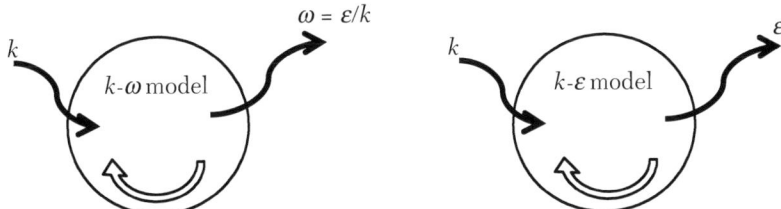

FIGURE 2.7: Comparison the k-ω and k-ε models based on the turbulence quantities involved.

The exact equation for ω can be derived using those of k and ε, and the following relationship, (see [20] for detailed derivation):

$$\frac{D\omega}{Dt} = \frac{D(\varepsilon/k)}{Dt} = \frac{1}{k}\frac{D\varepsilon}{Dt} - \frac{\omega}{k}\frac{Dk}{Dt} \tag{2.25}$$

Similar arguments, as those given for the k *and* ε equations, can be made to model the terms involved in the exact ω-equation to arrive at the CFD model equation. The resulting k-ω model has six coupled partial differential equations governing six variables: four equations for average velocity vector components and pressure and two equations for k and ω. In order to solve the system of PDEs, we need to have the numerical values of the model parameters which are defined for modeling different terms in the relevant equations. The model equations (2.26–2.29) for incompressible Newtonian fluids, as used in COMSOL, are the Wilcox revised two-equation k-ω model [20] for continuity, momentum, kinetic energy, and specific kinetic energy dissipation rate:

$$\overline{u}_{i,i} = 0 \tag{2.26}$$

$$\rho\left(\frac{\partial \overline{u}_i}{\partial t} + \overline{u}_j \frac{\partial \overline{u}_i}{\partial x_j}\right) = -\frac{\partial \overline{p}^*}{\partial x_i} + (\mu + \mu_T)\frac{\partial^2 \overline{u}_i}{\partial x_j \partial x_j} \tag{2.27}$$

$$\rho\left(\frac{\partial k}{\partial t} + \overline{u}_j \frac{\partial k}{\partial x_j}\right) = \frac{\partial}{\partial x_j}\left((\mu + \mu_T \sigma_k^*)\frac{\partial k}{\partial x_j}\right) + \overline{P}_k - \rho\beta_0^* k\omega \tag{2.28}$$

$$\rho\left(\frac{\partial \omega}{\partial t} + \overline{u}_j \frac{\partial \omega}{\partial x_j}\right) = \frac{\partial}{\partial x_j}\left((\mu + \mu_T \sigma_\omega)\frac{\partial \omega}{\partial x_j}\right) + \alpha\frac{\omega}{k}\overline{P}_k - \rho\beta_0\omega^2 \tag{2.29}$$

where, $\overline{P}_k = \mu_T \frac{\partial \overline{u}_i}{\partial x_j}(\overline{u}_{i,j} + \overline{u}_{j,i})$, $\mu_T = \rho\frac{k}{\omega}$, $\beta_0^* = 0.09$, $\beta_0 = 9/125$, $\alpha = 13/25$, $\sigma_k^* - \sigma_\omega = 0.5$.

Determination of constant parameters is based on a semi-empirical approach and using experimental results for decaying of isotropic turbulent kinetic energy and a turbulent boundary layer.

The corresponding k-ω equations (in vector notation) used in COMSOL are shown in Figure 2.8. For comparison, readers can identify equivalent terms in equations and benefit from the discussions, given in previous sections as well as this section for their physical interpretations and corresponding assumptions when using this model. For converting tensor notation to vector notation (or vice versa) please see [41].

Several versions of the k-ω model have been developed [20], most notably by Menter [62], [63], and [64], who in a series of publications suggested a modified version using the merits of both the k-ω and k-ε models. This model, known as the SST turbulence model, will be discussed in the next section.

The following points are related to analysis of the k-ω model and would be useful for application of the k-ω model and in comparison to the k-ε model.

1. Contrary to the k-ε model, the k-ω model behaves relatively much better close to a solid wall. As the value of $k \to 0$ close to the wall the value of $\omega \to \infty$, but this can be resolved by applying a wall function for the wall region. One should note

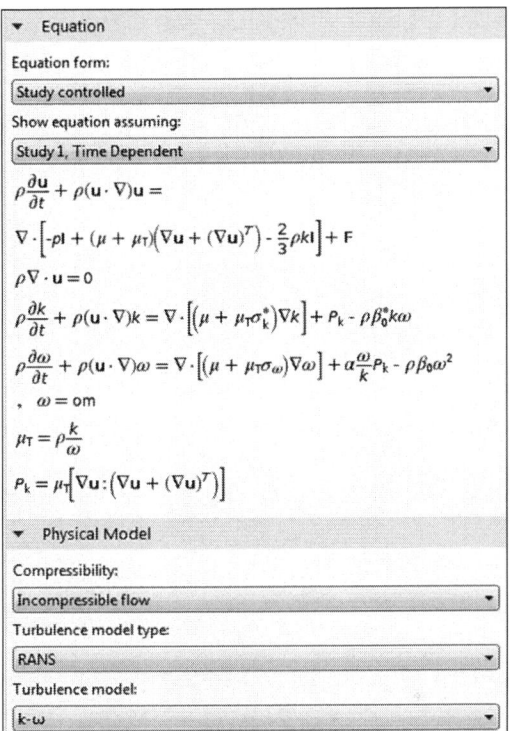

FIGURE 2.8: COMSOL k-ω model equations for incompressible fluids.

that by using relations for μ_T and $\overline{P}_k$, the production term $\left(\sim \dfrac{\omega}{k}\right)$ in the ω-equation (2.29) can be written as

$$\alpha \frac{\omega}{k} \overline{P}_k = \alpha \rho \frac{\partial \overline{u}_i}{\partial x_j}\left(\overline{u}_{i,j} + \overline{u}_{j,i}\right) \tag{2.30}$$

Therefore, there is no singularity for this term at the wall, where $k = 0$. In general k-ω model equations can be integrated up to the solid wall and predict a reasonable mean velocity profile. But the prediction of the model close to the wall, actually in the viscous sub-layer, is poor. For this reason in COMSOL a wall function is suggested and used with this model. We will discuss the topic of wall functions in a further section.

2. Realizability, and mixing-length constraints as well as local turbulence isotropy and equilibrium assumptions are

considered in this model in COMSOL, similar to the k-ε model, as discussed previously.

3. The equations for the k-ω model are non-linear and coupled, k-equation (2.28) and ω-equation (2.29) are highly coupled with each other and relatively lightly with the momentum conservation equation (2.27). This requires proper numerical methods in application for integrating these equations which is resolved in COMSOL using proper solvers.

4. The model predicts more reasonable results for flow separation close to a wall, as compared to the k-ε model. However, the k-ω model is more sensitive to free stream inlet boundary conditions. This can be analyzed using the relation for eddy viscosity, $\mu_T = \rho \dfrac{k}{\omega}$. For free stream flow, like external aerodynamic flows, the values of both k and ω tend to zero. Hence the eddy viscosity becomes indeterminate and a small non-zero value should be assumed for ω, which could affect the accuracy of the final modeling results.

Initial and Boundary Conditions: Similar to the k-ε model, boundary conditions for velocity and pressure are relatively straight forward. But for k and ω it is less clear for example, at the inflow and outflow boundaries their values are not known. For the symmetry plane, free surface, and outlet flow Neumann type boundary conditions (i.e. $\dfrac{\partial k}{\partial n} = 0$ and $\dfrac{\partial \omega}{\partial n} = 0$) could be used. For the inflow boundary, turbulence intensity I_T and length scale L_T are used, as shown in Figure 2.9, similar to what was discussed in previous section for the k-ε model. The wall function used in COMSOL for ω is

$$\omega_w = \frac{\rho k}{\mathcal{K} \mu \delta_w^+},$$ where $\mathcal{K} = 0.41$ is von-Karman constant.

Initial values are set as default values for k, similar to k-ε model, except for ω. For details, readers are referred to the COMSOL manual [55]. In COMSOL default initial conditions for both steady state and transient turbulent flows are set.

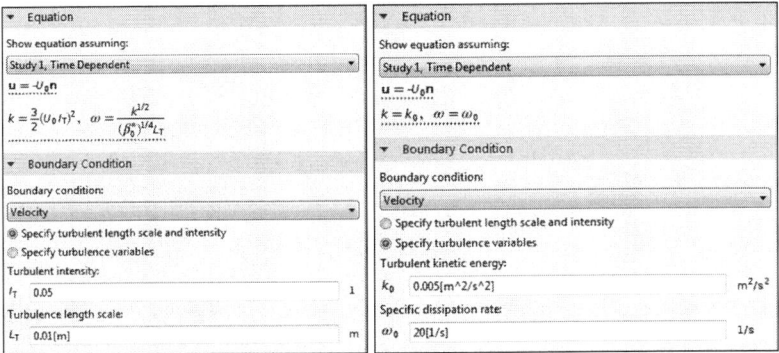

FIGURE 2.9: Typical boundary conditions for turbulent inflow in the COMSOL k-ω model.

SHEAR STRESS TRANSPORT k-ω MODEL

In this section, we summarize all the equations for the Shear Stress Transport (SST) model and compare them with those available in the COMSOL CFD module [55]. We also discuss general aspects of boundary conditions for the extra variables involved, k and ω, as well as the related constraints and types of turbulent flows for which the SST model could be applied with acceptable engineering 'accuracy.' In Chapter 4, we apply the COMSOL SST model and discuss details of some problem examples.

The SST model is actually a 'hybridized' model which combines the strengths and superior behaviors of the k-ε and k-ω models. As discussed previously, in general, k-ε model prediction is reasonable for flow regions away from the wall and insensitive to initial parameters of the main free stream flow, whereas k-ω behaves better close to the wall and for example, flows with adverse pressure gradient in a separated boundary layer region. Therefore, it seems useful to combine the advantages of these two models into a unified hybrid model. Menter [65], [63] did this and the result is the SST model, with several versions [62] for this model. He basically transformed the k-ε model to a modified k-ω model. He used the k-equation as it appears in the k-ε model (after some modification to the production term) and inserted $\varepsilon = k\omega$ into the ε-equation. For the latter, the resulting equation has an extra term which appears in the ω-equation of the SST model as compared to that of the *k-ω* model. A blending function is used for gradual change from the *k-ω* model in the region close to the wall in the boundary layer to a version of the *k-ε* model in the region far from the wall. Obviously, as a result of this treatment, the SST model does not require application of wall functions. The SST model also considers a modified formulation for turbulent eddy viscosity which accounts for transport of the principle turbulent shear stresses, hence the name SST model. Finally, this model gives, in general, more accurate results for the separated boundary layer, flow under adverse pressure gradient, flow around airfoils, and turbulent kinetic energy in stagnation regions [40].

The resulting SST model has six coupled partial differential equations governing six variables: four equations for average velocity vector components and pressure and two equations for *k* and *ω*. The model equations (2.31–2.34) for incompressible Newtonian fluids, as used in COMSOL, are Menter's SST model [55] for continuity, momentum, kinetic energy, and specific energy dissipation rate:

$$\overline{u}_{i,i} = 0 \tag{2.31}$$

$$\rho\left(\frac{\partial \overline{u}_i}{\partial t} + \overline{u}_j \frac{\partial \overline{u}_i}{\partial x_j}\right) = -\frac{\partial \overline{p}^*}{\partial x_i} + \left(\mu + \mu_T\right)\frac{\partial^2 \overline{u}_i}{\partial x_j \partial x_j} \tag{2.32}$$

$$\rho\left(\frac{\partial k}{\partial t} + \overline{u}_j \frac{\partial k}{\partial x_j}\right) = \frac{\partial}{\partial x_j}\left(\left(\mu + \mu_T\sigma_k\right)\frac{\partial k}{\partial x_j}\right) + P - \rho\beta_0^* k\omega \tag{2.33}$$

$$\rho\left(\frac{\partial \omega}{\partial t} + \overline{u}_j \frac{\partial \omega}{\partial x_j}\right) = \frac{\partial}{\partial x_j}\left(\left(\mu + \mu_T\sigma_\omega\right)\frac{\partial \omega}{\partial x_j}\right)$$

$$+ \rho\frac{\gamma}{\mu_T}P - \rho\beta\omega^2 + 2(1-F_1)\frac{\rho\sigma_{\omega 2}}{\omega}\frac{\partial \omega}{\partial x_j}\frac{\partial k}{\partial x_j} \tag{2.34}$$

The last term on the right-hand side of the *ω*-equation (2.34) is the new term which appears in the SST model, due to replacing *ε* = *kω* in the *ε*-equation of the *k-ε* model. As seen, this term involves an interpolation (or blending) function, F_1 (as defined in the COMSOL manual), which is used for interpolating constants $\beta, \gamma, \sigma_k, \sigma_\omega$, as

$$\beta = F_1\beta_1 + (1-F_1)\beta_2$$

$$\gamma = F_1\gamma_1 + (1-F_1)\gamma_2$$

$$\sigma_k = F_1\sigma_{k1} + (1-F_1)\sigma_{k2}$$

$$\sigma_\omega = F_1\sigma_{\omega 1} + (1-F_1)\sigma_{\omega 2} \tag{2.35}$$

where β_1 = 0.075, β_2 = 0.0828, γ_1 = 5/9, γ_2 = 0.44, σ_{k1} = 0.85, σ_{k2} = 1, $\sigma_{\omega1}$ = 0.5, and $\sigma_{\omega2}$ = 0.856. Eddy viscosity is defined as $\mu_T = \rho\dfrac{a_1 k}{\max(a_1\omega,|\overline{S}|F_2)}$ The function $F2$ is as defined in the COMSOL manual [55], a_1 = 0.31, and $|\overline{S}|$ is the magnitude of the mean strain-rate tensor, or $|\overline{S}| = \sqrt{2\overline{S}_{ij}\overline{S}_{ij}}$, where $\overline{S}_{ij} = \dfrac{1}{2}(\overline{u}_{i,j} + \overline{u}_{j,i})$. And finally $P = \min(\overline{P}_k, 10\beta_0^*\rho\omega k)$ where $\overline{P}_k$ is as defined previously for the k-ω model (i.e. $\overline{P}_k = \mu_T\dfrac{\partial \overline{u}_i}{\partial x_j}(\overline{u}_{i,j} + \overline{u}_{j,i})$) and β_0^* = 0.09.

The corresponding SST model equations (in vector notation) used in COMSOL are shown in Figure 2.10. For comparison, readers can identify equivalent terms in these equations and benefit from the discussions given in this section for their physical interpretations and corresponding assumptions when using this model. For converting tensor notation to vector notation (or vice versa) see [41].

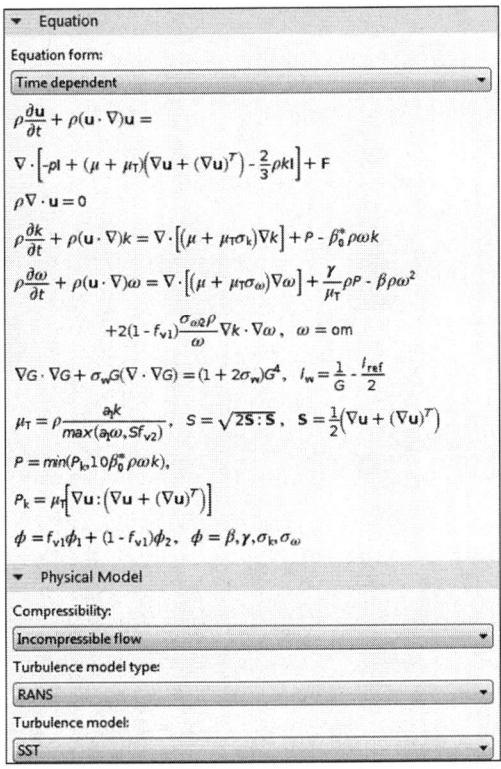

FIGURE 2.10: COMSOL SST model equations for incompressible fluids.

Following discussions related to analysis of the SST model seems useful, in comparison to the k-ε and k-ω models.

1. There are three new features in the SST model as compared to k-ε and k-ω models; (1) the introduction of a blending function F_1, (2) new cross-diffusion term in the ω-equation (2.34), and (3) new definition of turbulent eddy viscosity.

2. Blending function F_1 is a function of the ratio of the turbulence length scale, $\sqrt{k}/\omega$ over distance ℓ_w from the wall and turbulence Reynolds number, i.e. $\omega \ell_w^2/\nu$. In functional form, we can write

$$F_1 = F_1\left(\frac{\sqrt{k}}{\omega\,\ell_w}, \frac{\omega \ell_w^2}{\nu}\right) \tag{2.36}$$

This function is zero at the wall and asymptotes to unity far from the wall towards the main stream. In other words, we have

$$F_1 = \begin{cases} \to 0, & \text{approaching the wall} \\ \to 1, & \text{away from the wall} \end{cases}$$

Therefore, when considering the cross-diffusion term, i.e.

$2(1 - F_1)\dfrac{\rho\sigma_{\omega 2}}{\omega}\dfrac{\partial\omega}{\partial x_j}\dfrac{\partial k}{\partial x_j}$, we can conclude that the SST model behaves like the k-ω model close to the wall region, since $F_1 = 0$, and like k-ε model in the free stream or far from the wall region, since $F_1 = 1$ where the cross-diffusion term is null.

3. The new definition for eddy viscosity is based on Townsends' and Bradshaw's [66] assumption that the Reynolds shear stress, in a boundary layer, is proportional to the turbulent kinetic energy. In the SST model this assumption is used for limiting the eddy viscosity in regions where production of turbulent kinetic energy exceeds its dissipation.

Initial and Boundary Conditions: similar to the k-ω model, boundary conditions for velocity and pressure are defined at the wall and the inflow and outflow. For symmetry plane, free surface, and outlet flow Neumann type boundary conditions (i.e. $\dfrac{\partial k}{\partial n} = 0$ and

$\frac{\partial \omega}{\partial n} = 0$) could be used. For inflow boundary, turbulence intensity I_T and length scale L_T are used (see Table 2.2),as discussed for the k-ε model. For wall boundary condition, the no-slip condition, i.e. $\bar{u}_i = 0$ and $k = 0$, is applied. The boundary condition for ω is applied such to avoid singularity. Therefore the SST model equations are integrated up to the cells adjacent to the wall and the values of ω are calculated using a prescribed function, $\lim\limits_{\ell_w \to 0} \omega = \frac{6\mu}{\rho\beta_1\ell_w^2}$, as shown in Figure 2.11. Initial values are set as default values for k, similar to the k-ε model, except for ω. For details, readers are referred to the COMSOL manual [55].

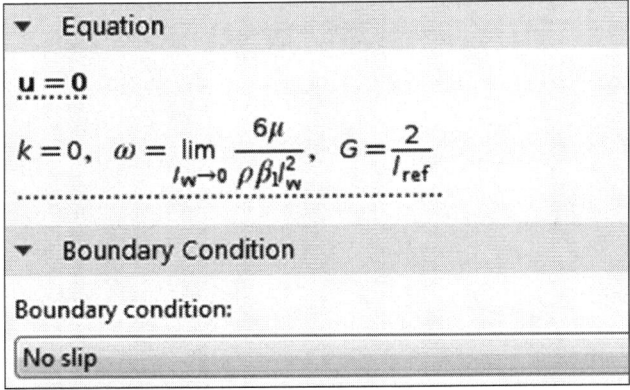

FIGURE 2.11: Boundary conditions at solid wall for SST model.

Overall, the SST model, in COMSOL, includes realizability and low-Reynolds number constraints, which means that this model is capable of modeling the flow all the way down to the wall. However, the SST model depends on the distance to the closest wall, therefore, the interface in COMSOL includes a wall distance equation. As shown in Figure 2.12, the Wall Distance Initialization study step is used. This first step is dedicated to solving for the reciprocal wall distance, i.e. variable G as shown in Figure 2.11. The distance determined in the initialization step is the distance to the closest wall. Convergence to the solution is not always quick, hence sometimes an initial solution could help, for example using a solution obtained by the k-ε model.

The COMSOL SST interface can be used for stationary and time-dependent analysis.

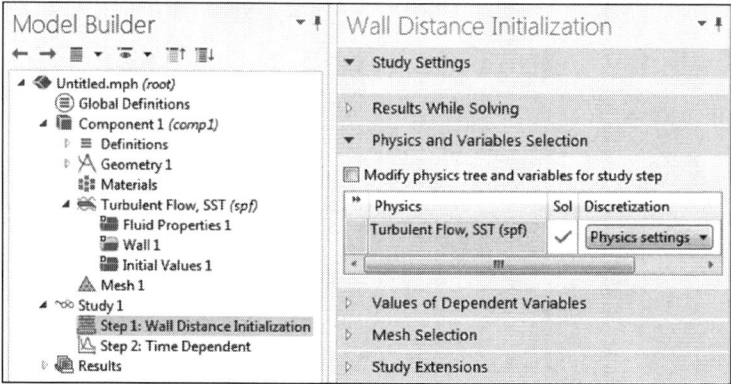

FIGURE 2.12: Wall distance initialization step as defined in SST model.

WALL FUNCTIONS

Real fluids stick to solid walls due to viscosity, for example consider flow in a pipe or a channel. Therefore, fluid velocity is zero at the stationary wall. To be more precise, the tangential velocity component is zero, and the normal component should satisfy the kinematic boundary condition of the wall, which is zero for a stationary solid wall. For a fully-developed turbulent flow, over a plate for example, the normalized velocity profile $u/\bar{u}_\infty$, parallel to the plate, and turbulent intensity are schematically shown in Figure 2.13.

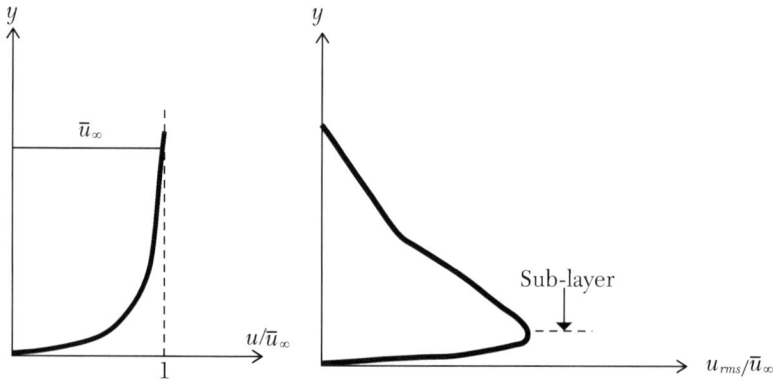

FIGURE 2.13: A typical velocity profile for turbulent velocity and turbulent intensity over a plate.

The velocity gradient $\frac{\partial \overline{u}}{\partial y}$, decreases as moving away from the wall or $\frac{\partial}{\partial y}\left(\frac{\partial \overline{u}}{\partial y}\right) < 0$. Shear stress is proportional to the velocity gradient through viscosity μ. At the wall, shear stress is $\tau_w = \mu \left.\frac{\partial \overline{u}}{\partial y}\right|_w$. As seen in Figures 2.13 and 2.14, the velocity profile indicates at least two length scales associated with the region close to the wall and the region with 'uniform' velocity profile away from the wall. A boundary layer thickness δ, is defined as the distance from the wall up to where the velocity is about 99% of its main stream value, or $\overline{u} = 0.99U_\infty$. Inside the boundary layer and very close to the wall there exists a so-called *viscous sublayer*, where velocity varies linearly with respect to normal distance from the wall, y. In this layer turbulence does not exist, as in the main flow, and viscous effect dominates or is at least comparable to inertia related effects. Now, away from the wall, in the so-called *outer region*, velocity is nearly constant and does not vary, or has very small variations, across the flow. The overall variation of velocity profile creates a challenge for turbulence models, at most, to find a reasonable solution for the averaged velocity profile both in the outer region and inside the boundary layer, including the viscous sub-layer. To overcome this challenge, two types of solutions are usually implemented: (1) either we have to modify the turbulence model considered so we can integrate all related model equations up to the wall (so-called 'Low-Reynolds' models), or (2) we find a function that can approximate the profile inside the boundary layer and then match the velocity profile to the more uniform one outside the boundary layer, i.e. bridging the velocity profile from the wall to the main flow. These types of functions are called *wall functions*. There exist several versions of wall functions including log-law, power-law (sometimes referred to as 1/7th power-law [67]), Spalding's law of the wall [68]. In COMSOL the so-called 'standard' wall function is used, hence we focus on this version for our discussion in this book.

In general, there are four flow regimes for a turbulent flow near a wall. As mentioned, adjacent to the wall there exists a viscous sub-layer (or laminar sub-layer) and above it there is the so called *buffer layer*. In the buffer layer flow starts to go through transition towards turbulence. Above the buffer layer, there exists a layer called *log-layer*, where the velocity is proportional to the logarithm of the distance from the wall. The thickness of log-layer is roughly orders of magnitude (about 100 times) larger than those of viscous and buffer layers, in total.

We continue this discussion with special attention paid to the behaviors of quantities like k, ε and ω near solid walls and associated turbulence models to have a brief analysis of the wall functions. We already mentioned that there needs to be a minimum of two length scales associated with the velocity profile for flow bounded to a wall. Let's define a problem-level length scale, like radius of a pipe, for example R, associated with the flow in the outer region. To get a length scale for the sub-layer, where we have a linear variation for velocity versus distance from the wall, we need a velocity scale $u_w \sim \sqrt{\dfrac{\tau_w}{\rho}}$, so-called friction velocity. Then the length scale for sub-layer would be v/u_w (since viscosity dominates in sub-layer and it is proportional to 'velocity' times 'length,' using dimensional analysis). The length scale for the region between the sub-layer and the outer layer, over which the two velocity profiles should match, has to be very small relative to R and very large as compared to v/u_w [19], or

$$v/u_w \ll y \ll R \tag{2.37}$$

In other words, in a region defined by y distances from the wall we should find a function that 'merges' the velocity profile in the sub-layer to that of the main stream. For this purpose, we use an argument similar to eddy viscosity to relate the Reynolds shear stress to the gradient of the mean velocity, as

$$-\overline{u'v'} = \frac{\tau_w}{\rho} = v_e \frac{\partial \overline{u}}{\partial y} \sim u_w^2 \tag{2.38}$$

But $v_e = A u_w y$, using dimensional analysis or inequality (2.37). Where, A is constant of proportionality. After plugging in for v_e into equation (2.38), we have (after replacing partial differentiation with ordinary one)

$$\frac{d\overline{u}}{dy} = C u_w/y \tag{2.39}$$

where, C is a constant. It is useful to define a velocity scale and length scale for this region, as well. Since, this region spreads over distances, given by y, from the wall, associates with two velocities; $\bar{u}$ from the outer region and u_w from the sublayer. The velocity scale is taken to be u_w, which gives the dimensionless velocity $u^+ = \dfrac{\bar{u}}{u_w}$. Similarly the length scale for this region is ν/u_w, with dimensionless distances from the wall, or a local Reynolds number, being as $y^+ = \dfrac{y u_w}{\nu}$. By expressing equation (2.39) in terms of dimensionless variables u^+ and y^+, we have

$$\frac{du^+}{dy^+} = C/y^+ \qquad (2.40)$$

After integration, immediately we have:

$$u^+ = C \ln y^+ + B \qquad (2.41)$$

Equation (2.41) is the well-known log-law of wall, which provides a function for velocity to match the inner sub-layer to the outer layer and is extensively verified experimentally [69]. Both Constants (i.e. C and B) can be determined by experimental data and $C = 1/\kappa$, where $\kappa \cong 0.41$ is von Kármán constant and $B \cong 5.5$ (a range of values from 4.9 to 5.5 is given for B).Because of the logarithmic functional form of this function, the corresponding layer, in the y range distances from the wall, is called log-layer ($30 \lesssim y^+ \lesssim 1000$). This layer is above the sublayer ($0 \lesssim y^+ \lesssim 5$) (with the existence of a possible buffer layer in between, $5 \lesssim y^+ \lesssim 30$) and below the outer layer. For further detail on the boundary layer, readers may refer to Schlichting [70] and [46]. Figure 2.14 shows schematics of different layers in a turbulent boundary layer.

By applying log-law for inner region close to a wall, we can save computational time when modeling turbulence, since very fine mesh resolution is not required in this region as a consequence. In this way velocity profile can be calculated using the log-law and should be applied as boundary conditions for the turbulence model variables at the 'border' of the outer layer. There are some limitations to the application of log-law due to its validity for basic assumption of fully-developed turbulent boundary layer. Therefore, as mentioned previously in this section, several improvements have been suggested with the introduction of different types of wall functions, such as Power low and Spalding's law of the wall [68]. For related

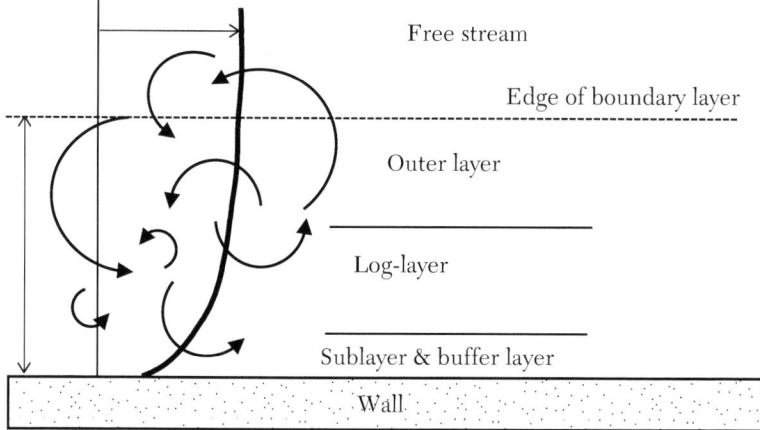

FIGURE 2.14: Sketch of typical turbulent boundary layer and associated layers in relation to main free stream.

detail discussions, see [22] and [71] for non-equilibrium wall functions, and [46] for comparison of power-law versus log-law.

In COMSOL, log-law is used for wall functions up to a distance of $y^+ = 11.06$ for the k-ε model and k-ω model, (y^+ is designated by δ_w^+ in COMSOL).

Now we discuss the boundary conditions, in relation to wall function, for models k-ε and k-ω. Selection of a wall function is required for either of these two models in COMSOL. The discussion rests on the assumption that in a fully-developed boundary layer close to a wall, i.e. in log-layer, the turbulence production and dissipation are much larger than turbulence diffusion and convection [27]. Therefore the momentum equation reads $\dfrac{\partial}{\partial y}\left(v\dfrac{\partial \bar{u}}{\partial y}\right) = 0$. By applying the eddy viscosity, which varies linearly from the wall distance in the log-layer, and after some manipulations (see [21]) we obtain the following equations for the k-ε model in a log-layer

$$\bar{u} = \frac{u_w}{\kappa}\ln y^+ + B$$

$$k = \frac{u_w^2}{\sqrt{C_\mu}} \tag{2.42}$$

$$\varepsilon = \frac{u_w^3}{\kappa y}$$

Using COMSOL's notation $(y = \delta_w, y^+ = \delta_w^+ = \dfrac{\rho y u_w}{\mu}, \kappa = \kappa_V)$, the

ε-equation reduces to $\varepsilon = \dfrac{\rho k^2}{\kappa_V \delta_w^+ \mu}$. This is exactly (besides, C_μ) what

we have for wall functions in the k-ε model, as shown in Figure 2.15,

under Study1, Wall Distance Initialization section.

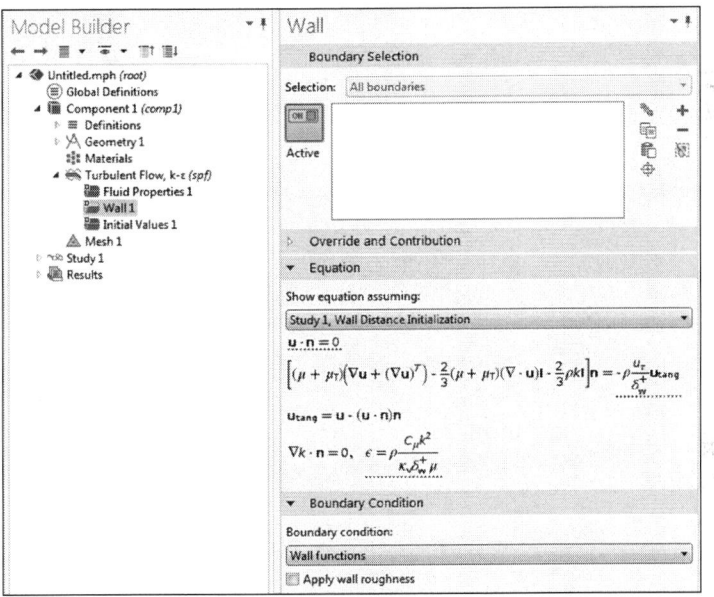

FIGURE 2.15: Wall boundary conditions for k-ε model.

Similarly, equations for k-ω model in a log-layer are

$$\bar{u} = \frac{u_w}{\kappa} \ln y^+ + B$$

$$k = \frac{u_w^2}{\sqrt{\beta_0^*}} \qquad (2.43)$$

$$\omega = \frac{u_w}{\kappa y \sqrt{\beta_0^*}}$$

Using COMSOL's notation $(y = \delta_w, y^+ = \delta_w^+ = \dfrac{\rho y u_w}{\mu}, \kappa = \kappa_V)$, the

ε-equation reduces to $\omega = \dfrac{\rho k}{\kappa_V \delta_w^+ \mu}$. This is exactly what we have for

wall functions in k-ω model, as shown in Figure 2.16, under Study1,
Wall Distance Initialization section.

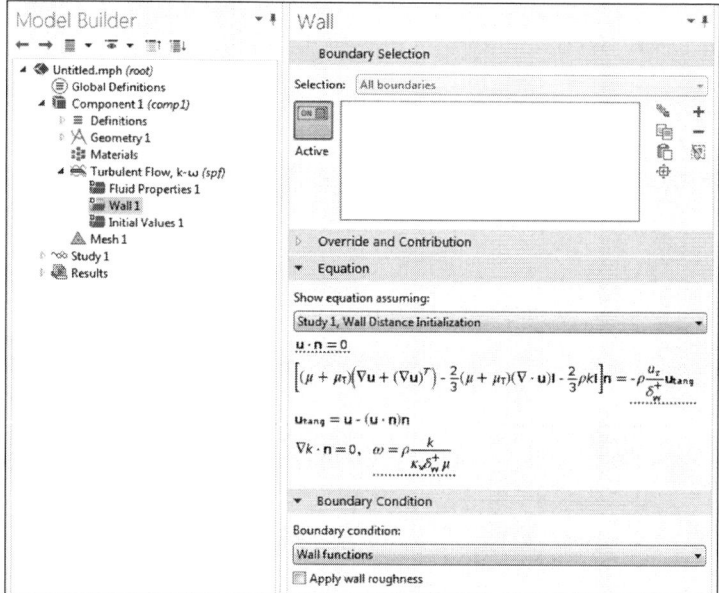

FIGURE 2.16: Wall boundary conditions for *k-ω* model.

Readers are referred to the previous sections related to k-ε and k-ω models for comparing full model equations, in connection with wall functions application.

LOW-Re k-ε MODEL

In this section, we summarize all the equations for Low-Reynolds-number k-ε model and compare them with those available in the COMSOL CFD module [55]. We also discuss general aspects of boundary conditions for extra variables involved as well as the related constraints and types of turbulent flows to which the low-Re k-ε model could be applied with acceptable engineering 'accuracy.'

This model is basically a modified and extended version of the k-ε model. In the previous section, we discussed wall functions that are used for modeling turbulence flow close to a solid wall in conjunction with a turbulence model, for example the k-ε model. As mentioned, wall functions are approximation for treatment of flow in a boundary layer close to a wall, with the purpose of saving computer time and power. However, for some turbulent flows it is desirable to have a more 'exact' solution of the boundary layer flow. For example, in external flow around airfoils, cars, conjugate heat transfer, lift and drag forces calculations, etc. Therefore, it is very important for engineering applications to have a turbulence model which can model the flow close to a solid wall. One remedy to resolve this challenge is to have a very fine mesh close to the wall to capture the turbulence details. This will increase the computational costs, as well as brings up the conflicting assumptions; that is the very high Reynolds number for a turbulence model, like the k-ε model, versus the viscous-effect domination in the sub-layer $y^+ \lesssim 5$. In other words, we have to modify the k-ε model equations in a way that we can integrate them up to the wall. The result is the *Low-Re* k-ε model. It should be mentioned that the name Low-Re is given to the k-ε model, or in general any other turbulence model, to indicate that the model can handle the flow modeling in regions with a low-Reynolds number, for example close to a solid wall. Another important point is that the Reynolds number which is referred to here is the local turbulence Reynolds number (not the global Reynolds number based on the problem velocity and length scale) based on local turbulence velocity and length scales or $Re_T = \dfrac{U\ell}{\nu}$. This Reynolds number varies throughout the flow/modeling domain and goes to zero when a solid wall is approached. It should be mentioned that the local Reynolds number can be written as $Re_T = \dfrac{\nu_T}{\nu}$. and could be also interpreted as the ratio of turbulent kinematic viscosity ν_T which is a property of the flow, and fluid kinematic viscosity ν, which is a property of the fluid.

There exist several versions of the Low-Re k-ε model, for example the AKN model [72], the CHC model [73], the LS model [74], the YS model [75], and Patel et al. [76]. COMSOL uses the AKN model, which has also shown superior performance for turbulent flow around underwater vehicles [77]. For understanding the model equations we mention the approach used (for example in the references mentioned above) for their derivations.

General approach [78] to derive the Low-Re k-ε model equations is to consider the behavior of terms involved in exact equations for k and ε as $y \to 0$, i.e. a wall is approached, and compare them with the corresponding modeled equations terms. This comparison leads to definition of so-called *damping function* f_μ, which is used to modify the turbulent eddy viscosity, $\mu_T \sim f_\mu \dfrac{k^2}{\varepsilon}$ and consequently modifies the production and diffusion terms in the k-ε model. Damping function behaves like $f_\mu = \mathcal{O}(1/y)$ when $y \to 0$, and $f_\mu \to 1$ when $y^+ \geq 50$. In COMSOL two damping functions are used, designated by f_μ and f_ε. The latter is used to modify the dissipation term in ε-equation. The Low-Re k-ε model equations (2.44–2.47), for an incompressible Newtonian fluid, for continuity, momentum, kinetic energy, and energy dissipation, are:

$$\overline{u}_{i,i} = 0 \tag{2.44}$$

$$\rho\left(\frac{\partial \overline{u}_i}{\partial t} + \overline{u}_j \frac{\partial \overline{u}_i}{\partial x_j}\right) = -\frac{\partial \overline{p}^*}{\partial x_i} + (\mu + \mu_T)\frac{\partial^2 \overline{u}_i}{\partial x_j \partial x_j} \tag{2.45}$$

$$\rho\left(\frac{\partial k}{\partial t} + \overline{u}_j \frac{\partial k}{\partial x_j}\right) = \frac{\partial}{\partial x_j}\left(\left(\mu + \frac{\mu_T}{\sigma_k}\right)\frac{\partial k}{\partial x_j}\right) + \overline{P}_k - \rho\varepsilon \tag{2.46}$$

$$\rho\left(\frac{\partial \varepsilon}{\partial t} + \overline{u}_j \frac{\partial \varepsilon}{\partial x_j}\right) = \frac{\partial}{\partial x_j}\left(\left(\mu + \frac{\mu_T}{\sigma_\varepsilon}\right)\frac{\partial \varepsilon}{\partial x_j}\right) + C_{\varepsilon 1}\frac{\varepsilon}{k}\overline{P}_k - C_{\varepsilon 2}\rho\frac{\varepsilon^2}{k}f_\varepsilon \tag{2.47}$$

Where $\overline{p}^* = \overline{p} + \dfrac{2}{3}\rho k$, $\overline{P}_k = \mu_T \dfrac{\partial \overline{u}_i}{\partial x_j}\left(\overline{u}_{i,j} + \overline{u}_{j,i}\right)$, $\mu_T = \rho f_\mu C_\mu \dfrac{k^2}{\varepsilon}$, $C_\mu = 0.09$, $C_{\varepsilon 1} = 1.5$, $C_{\varepsilon 2} = 1.9$, $\sigma_k = 1.4$, and $\sigma_\varepsilon = 1.5$. The damping functions are calculated automatically in the background in COMSOL, using a wall distance variable $_{lw}$ which is provided in the Wall Distance Initialization interface.

The corresponding Low-Re k-ε equations (in vector notation) used in COMSOL are shown in Figure 2.17. By comparison, readers can identify equivalent terms in equations and benefit from the discussions in this section. For converting tensor notation to vector notation (or vice versa) see [41]. Discussions given in the previous section, under the k-ε model, for physical interpretations of terms involved in the model equations are valid when damping functions are included, as well. Readers are encouraged to refresh their readings from the k-ε model in connections with Low-Re k-ε model equations.

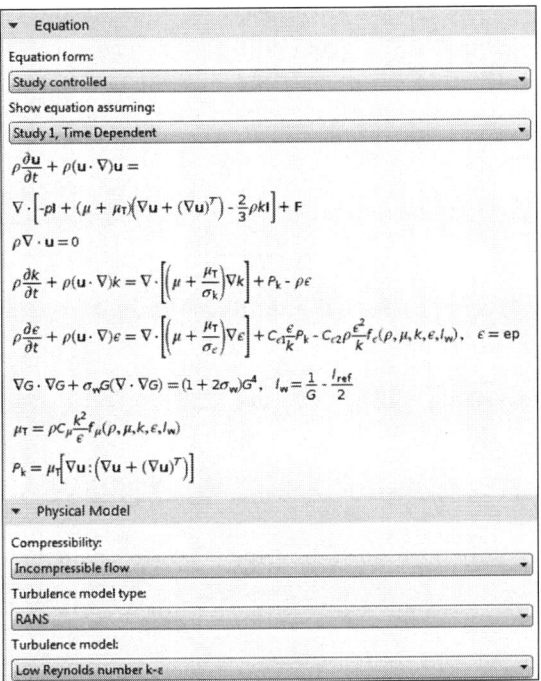

FIGURE 2.17: COMSOL Low-Re k-ε model equations for incompressible fluids.

At this point, it seems useful to have a closer look at the damping functions and their functional forms. Both functions are functions of density, viscosity, turbulent kinetic energy and its dissipation rate, and distance to the wall. Mathematically, we have:

$$f_\mu = \text{Function } (\rho, \mu, k, \varepsilon, y)$$
$$f_\varepsilon = \text{Function } (\rho, \mu, k, \varepsilon, y)$$

By using Kolmogorov velocity scale $u_\varepsilon = (\mu \varepsilon/\rho)^{1/4}$ we can define a dimensionless wall distance as $y^k = (\rho u_\varepsilon y)/\mu$ and a local turbulence Reynolds number as $Re_T = \dfrac{v_T}{v} = \dfrac{\rho k^2}{\varepsilon \mu}$. Using these non-dimensional variables, the explicit functional forms of damping functions read [72]:

$$f_\mu = \left[1 - \exp\left(\frac{-y^k}{14} \right) \right]^2 \left\{ 1 + \frac{5}{Re_T^{3/4}} \exp\left[-\left(\frac{Re_T}{200} \right)^2 \right] \right\} \qquad (2.48)$$

$$f_\varepsilon = \left[1 - \exp\left(\frac{-y^k}{3.1} \right) \right]^2 \left\{ 1 - 0.3 \, \exp\left[-\left(\frac{Re_T}{6.5} \right)^2 \right] \right\} \qquad (2.49)$$

In COMSOL [55], distance from the wall is designated as l_w, hence $y = l_w$ and dimensionless wall distance as l^* hence $y^k = l^*$.

Boundary and Initial Conditions: no-slip boundary condition could be applied for velocity and hence, $k = 0$, at the wall. Boundary condition for e is not defined at the wall, rather at the cell adjacent to the wall at a distance of $2\dfrac{\mu k}{\rho l_w^2}$. These are default values at the wall and set automatically in COMSOL. Boundary conditions at inlet are selected similar to what was discussed for the k-ε model, such as turbulent intensity and length scale. Initial values are similar to the k-ε model. However, it is recommended [55] to solve the problem at hand using the k-ε model and then use the results as initial values for the Low-Re k-ε model. This approach helps convergence of the numerical solution. According to the COMSOL manual, the procedure is as follows:

1. Solve the model using the k-ε model.

2. Change the turbulence model to a Low-Re k-ε model.

3. Add a new study using Stationary with Initialization and set Values of variables not solved for to Solution from the first study.

4. Solve for new study.

The Turbulent Flow, Low Re k-ε interface in COMSOL can be used for stationary and time-dependent analysis. The physics interface is suitable for incompressible flows and compressible flows at low Mach number flows (typically less than 0.3).

SPALART-ALLMARAS MODEL

In this section, we summarize all the equations for the Spalart-Allmaras (S-A) model and compare them with those available in the COMSOL CFD module [55]. We also discuss general aspects of boundary conditions for extra variables involved as well as the related constraints and types of turbulent flows for which the S-A model could be applied with acceptable engineering 'accuracy.' In

Chapter 4, we apply the COMSOL S-A model and discuss details of some problem examples.

The S-A model is a relatively recent turbulent model, developed by Spalart and Allmaras [79]. This model is a one-equation model type and uses a single conservation equation for a variable in order to calculate eddy viscosity, using Boussinesq's hypothesis. As mentioned in the previous sections, the objective of all RANS models, at least those using Boussinesq's hypothesis, is to calculate turbulent eddy viscosity using relevant turbulence scales. The S-A model approach is to shortcut this requirement, and it directly aims to calculate for eddy viscosisty. Of course, the requirement for determining a length scale for treatment of flow close to walls remains and the model depends on the distance to the closest wall.

There are many versions of the S-A model available, as listed in the NASA Turbulence Modeling Resources website [80]. The model is computationally economical with relatively good performance for aerodynamic type of applications involving flow around airfolis, wings, etc., external flows with mild flow separation, wall-bounded flows, and turbomachinery applications. The model has shown relatively good results for boundary layers with adverse pressure gradient, or positive pressure gradient in the direction of the flow. The model does not perform well for free shear flows. However, the model is robust and converges with reasonable coarse mesh. It is sometimes used for obatining initial values for more advanced turbulenece models, for example, the Low-Re k-ε model. This model is also a 'low-Re' and hence, does not need application of wall functions and it models the flow all the way down to the wall.

Highly-populated-parameter S-A model development includes a transport equation for a working variable, $\tilde{v}$, which is used for calculating eddy viscosity,as $\mu_T = \rho \tilde{v} f_{v1}$. The function f_{v1} is a damping function and $f_{v1} \to 1$ for high Reynolds numbers and $f_{v1} \to 0$ when a solid wall is approached. The variation of f_{v1} is favorable,due to high Reynolds numbers region $\tilde{v} = v_T$, which recovers the kinematic eddy viscosity. In COMSOL, a standard S-A model without trip terms [20] is available.

The model equations consist of five equations, governing five unknowns (i.e. $\overline{u}_i, \overline{p}, \tilde{v}$). For the length scale, $l_w = y$ is selected and is defined as the distance to the closest wall. The transport equation for, along with continuity and momentum equations are the S-A model equations (2.50–52) for Newtonian incompressible fluids, as follows:

$$\overline{u}_{i,i} = 0 \tag{2.50}$$

$$\rho\left(\frac{\partial \overline{u}_i}{\partial t} + \overline{u}_j\frac{\partial \overline{u}_i}{\partial x_j}\right) = -\frac{\partial \overline{p}^*}{\partial x_i} + (\mu + \mu_T)\frac{\partial^2 \overline{u}_i}{\partial x_j \partial x_j} \tag{2.51}$$

$$\frac{\partial \tilde{v}}{\partial t} + \overline{u}_j\frac{\partial \tilde{v}}{\partial x_j} = P_{\tilde{v}} - D_{\tilde{v}} + \frac{1}{\sigma_{\tilde{v}}}\left[\frac{\partial}{\partial x_j}\left((v + \tilde{v})\frac{\partial \tilde{v}}{\partial x_j}\right) + C_{b2}\frac{\partial \tilde{v}}{\partial x_i}\frac{\partial \tilde{v}}{\partial x_i}\right] \tag{2.52}$$

where $\overline{p}^* = \overline{p} + \frac{2}{3}\rho k$, $P_{\tilde{v}} = C_{b1}\tilde{S}\tilde{v}$, $D_{\tilde{v}} = C_{w1}f_w\left(\frac{\tilde{v}}{\kappa y}\right)^2$, $\mu_T = \rho\tilde{v}f_{v1}$,

$$\mu = \rho v, \quad f_{v1} = \frac{\chi^3}{\chi^3 + C_{v1}^3}, \quad \chi = \frac{\tilde{v}}{v}, \quad C_{w1} = \frac{C_{b1}}{\kappa^2} + \frac{1 + C_{b2}}{\sigma_{\tilde{v}}},$$

$$f_w = g\left[\frac{1 + C_{w3}^6}{g^6 + C_{w3}^6}\right]^{1/6}, \quad g = r + C_{w2}(r^6 - r), \quad r = \min\left(\frac{\tilde{v}}{\tilde{S}\kappa^2 y^2}, 10\right),$$

$f_{v2} = 1 - \frac{\chi}{1 + \chi f_{v1}}$, $C_{b1} = 0.1355$, $C_{b2} = 0.622$, $C_{v1} = 7.1$, $\sigma_{\tilde{v}} = 2/3$,

$C_{w2} = 0.3$, $C_{w3} = 2$, $\kappa = 0.41$, and $C_{Rot} = 2$.

The only parameter that remains to be defined is the magnitude of modified vorticity $\tilde{S}$. Recall rotation tensor (or asymmetric part of the velocity gradient) was defined as, $\xi_{ij} = \frac{1}{2}\left(\frac{\partial u_i}{\partial x_j} - \frac{\partial u_j}{\partial x_i}\right)$. Then $\tilde{S} = \xi + \frac{\tilde{v}f_{v2}}{\kappa^2 y^2}$, where $\xi = |\xi_{ij}|$ is the magnitude of vorticity (defined as Ω in COMSOL). The definition of $\tilde{S}$ is further modified to satisfy the requirement of $\tilde{S} > 0$ and no smaller than 30% of ξ, [80]. In COMSOL [55] this is satisfied by using a more complex relation as

$$\tilde{S} = \max\left[\left(\xi + C_{Rot}\min(0, S - \xi) + \frac{\tilde{v}f_{v2}}{\kappa^2 y^2}\right), 0.3\xi\right], \text{ where } S \text{ is the mag-}$$

nitude of mean rate of strain tensor.

The corresponding S-A model equations (in vector notation) used in COMSOL is shown in Figure 2.18. By comparison, readers can identify equivalent terms in equations and benefit from the discussions, given in this section. For converting tensor notation to vector notation (or vice versa) please see [41].

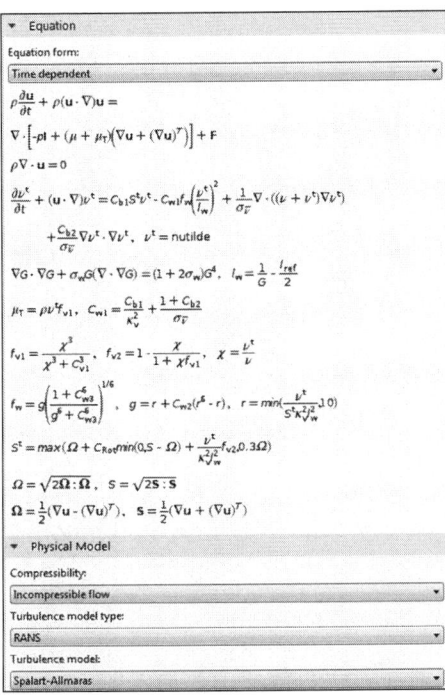

FIGURE 2.18: COMSOL Spalart-Allmaras model equations for incompressible fluids.

The author[12] discovered two errors/typos associated with the S-A model equations in COMSOL, which were reported to COMSOL and corrected in COMSOL, version number 4.4.0.248 and beyond. Readers should check their installed COMSOL software version number to make sure that they have the latest version. If the version number is older than 4.4.0.248, then an updated version should be downloaded or, otherwise, make corrections, for C_{w1} and r, according to their relations given under equation (2.52) or in Figure 2.18.

Boundary and Initial Conditions

Boundary conditions for velocity and $\tilde{v}$ are set to zero at the stationary solid wall. The boundary conditions at a symmetry plane for $\tilde{v}$ is of Neumann type, or $\dfrac{\partial \tilde{v}}{\partial n} = 0$ and for free stream $\bar{v}$ = (3 to

---

[12] Here, we would like to acknowledge the quick response and professional attention received from the COMSOL support team pertinent to this communication (March 19, 2014). The errors/typos are corrected in version number 4.4.0.248 and beyond.

5)v. The initial value for $\tilde{v}$ is set by default using a scale parameter of $5 \times 10^{-6}\,\text{m}^2/\text{s}$. For other variables, absolute scales like those used for the k-ε model apply [55].

Finally, we should mention that the Spalart-Allmaras interface in COMSOL can be used for simulating single-phase flows at high Reynolds numbers for incompressible flows, and compressible flows at low Mach number flows (typically less than 0.3). The interface can also be used for stationary and time-dependent analysis.

ALGEBRAIC yPLUS MODEL

In this section, we summarize the governing equations for the Algebraic yPlus model. This model is categorized under zero-equation group. It is a 'fast' and low-demanding model in terms of computer resources and useful for 2D flow without boundary layer separation or flow with mild curvature.

The yPlus turbulent model is based on Prantl's mixing length model ([67], [42]), which uses algebraic relation for calculating the eddy viscosity. Prantl's mixing-length theory is based on analogy borrowed from the kinetic theory of gases, that is, assuming a length scale for turbulent flow within which turbulent momentum exchange/transfer occurs. This length scale is called mixing-length, l_m. Using dimensional analysis for eddy viscosity, we can write $\mu_T = \rho l_m^2 \left| \dfrac{d\bar{u}}{dy} \right|$. By using this model we can calculate eddy viscosity using gradient of average velocity, once l_m is known. But the mixing length, or the length scale of turbulence, is not constant and, as discussed previously, turbulence has a range of length scales. In the yPlus model a non-linear algebraic equation is solved (as a function of position, or at each node pint of the finite element domain) for the dimensionless wall distance y^+, that is in turn used to calculate the eddy viscosity in its dimensionless form (see the **COMSOL 5** manual). The governing equations for this model are those given by equations (2.13 and 2.14), which are repeated here for convenience.

$$\rho\left(\frac{\partial \bar{u}_i}{\partial t} + {}_j \frac{\partial \bar{u}_i}{\partial x_j} \right) = -\frac{\partial \tilde{p}^{\circ}}{\partial x_i} + \left(\mu + \mu_T \right) \frac{\partial^2 \bar{u}_i}{\partial x_j \partial x_j} \qquad (2.13)$$

$$\overline{}_{i\,i} \tag{2.14}$$

The yPlus model requires defining a wall distance, which is automatically solved in COMSOL using the parameter l_{ref} (see the COMSOL Manual).

L-VEL MODEL

In this section, we summarize the governing equations for the L-VEL model. This model is categorized under zero-equation group and is an extension of the logarithmic law of the wall. It is a 'fast' and low-demanding model in terms of computer resources. It is usually used for turbulent flow modeling in a region close to a wall with multi-scale features, like electronic circuit board cooling design. It was developed by Agonafer et al. [53] and solves for local turbulent viscosity for a given distance from the wall. Once eddy viscosity is known at each node, then the governing equation for L-VEL or RANS, given by equations (2.13 and 2.14), can be solved.

As mentioned previously (see Wall Functions section), in the log-layer the dimensionless distance to the wall, $y^+ = \dfrac{y u_w}{v}$ and dimensionless velocity $u^+ = \dfrac{\overline{u}}{u_w}$ are related through a logarithmic relation, or the log law of the wall, as given by equation (2.41). We repeat these equations here for convenience, with its coefficient expressed explicitly.

$$u^+ = \frac{1}{\kappa}\ln y^+ + B \tag{2.41}$$

Spalding [68] expanded on this law by using a Taylor expansion of the y^+ as an exponential function of u^+ and including the linear sublayer, as well. In doing so, only five terms of the Taylor expansion was considered, which leads to the so-called Spalding's law of the wall, as

$$y^+ = u^+ + \frac{1}{E}\left(e^{\kappa u^+} \underbrace{-1 - \kappa u^+ - \frac{\left(\kappa u^+\right)^2}{2} - \frac{\left(\kappa u^+\right)^3}{6} - \frac{\left(\kappa u^+\right)^4}{24}}_{\text{from Taylor expansion of } e^{\kappa u^+}} \right) \tag{2.53}$$

Where constant $E = e^{\kappa B}$, is 8.6 (for $\kappa = 0.417$ and $B = 5.16$). By differentiating equation (2.53) a relationship for eddy viscosity can be derived, since shear stress is proportional to gradient of velocity, as given in equation (2.54):

$$v^+ = 1 + \frac{\kappa}{E}\left(e^{\kappa u^+} - 1 - \kappa u^+ - \frac{\left(\kappa u^+\right)^2}{2} - \frac{\left(\kappa u^+\right)^3}{6}\right) \qquad (2.54)$$

Where effective viscosity $v^+ = \dfrac{v_T}{v}$, or ratio of turbulent eddy viscosity over fluid kinematic viscosity. Using this equation, one can calculate the effective viscosity, once u^+ is known. This is achieved by defining a local Reynolds number $\widehat{Re} = u^+ y^+$, or

$$\widehat{Re} = u^+\left(u^+ + \frac{1}{E}\left(e^{\kappa u^+} - 1 - \kappa u^+ - \frac{(\kappa u^+)^2}{2} - \frac{(\kappa u^+)^3}{6} - \frac{(\kappa u^+)^4}{24}\right)\right) \qquad (2.55)$$

Using an iterative procedure, like Newton-Raphson, we can calculate u^+ from (2.55) at each node and use equation (2.54) to calculate v^+

The L-VEL model requires defining a wall distance, which is automatically solved in COMSOL using the parameter l_{ref} (see the COMSOL 5 Manual).

General Guideline for Turbulent Models Application

No turbulence model is universally accepted as the superior model. However, each model, based on its development and calibration, has merits and could be more suitable for certain types of flows than others. Before choosing a model it is recommended to calculate the flow Reynolds number and understand the flow physics and behavior. The following guideline might be helpful in choosing a 'suitable' turbulent model.

- The k-ε model with wall function for flow with no adverse pressure gradient or strong separation.

- If more accurate results (e.g. heat transfer, drag, lift) are needed for flow near the wall, use low-Re k-ε. In such a case, the k-ε model could be used for finding an initial solution to help the convergence.

- The k-ω model could be used, along with wall function when boundary layer separation and boundaries with flow curvature exist.

- The SST model is more comprehensive, which combines the strengths of k-ε and k-ω, without using a wall function. This model is relatively more demanding in terms of computer resources.

- The S-A model is a relatively simpler model and more suitable for flow around airfoil and blades, using separated flows rather than free-shear flows. This model is relatively less demanding in terms of computer resources compared to k-ε or k-ω models.

- For an initial solution the k-ε model could be used and then upgraded to k-ω, SST, or S-A models, if needed.

The choosing-by-elimination method could be more useful and less confusing when a turbulence model selection is in question. For example for an internal flow with no or mild curvature and no separation, either the k-ε or k-ω model could be used. When adverse pressure gradient of strong flow separations exist (e.g. flow in a diverging pipe) the k-ε model is not suitable. When turbulence close to a wall is important (e.g. heat transfer) the Low Re k-ε model would be more useful. For internal or external flow, SST is suitable while considering the cost of required computer resources and time. For flow around wings, airfoils, and when a fast and less expensive solution is sought, S-A could be used. In any case, users should possibly validate their results against published results, either experimental or numerical.

Table 2.3 summarizes, relatively, features and applications, but not limited to the turbulence models available in COMSOL.

TABLE 2.3: Features and applications of turbulence RANS models available in COMSOL.

Model	Advantages/ flows	Shortcomings/ flows	Wall function need	Computational cost/iteration (rank: 1 is least)
k-ε	Well established and validated, common, suitable for fully turbulent flows	Flow separation, streamline curvature	Yes	3
k-ω	Close-to-wall, separation, flow with curvature, complex boundary layer flows	Sensitive to values at free boundaries	Yes	4
SST	Near wall and free stream, more comprehensive	Computer time and resource	No	6
Low-Re k-ε	More detailed near-wall turbulence and transfer	Computer time and resource	No	5
S-A	Airfoil, wings, boundary layer with pressure gradient	Free-shear flow, 3D flow	No	2
yPlus	Low demand on computer power, quick-fast results, simplicity, 2D attached flow, economical	Complex flow with separation, non-local turbulence effects, single length scale, lack of turbulence non-local effects	No	1
L-VEL	Low demand on computer power, quick-fast results, simplicity, 2D/3D attached flow, economical	Complex flow with separation, non-local turbulence effects, single-length scale, lack of turbulence non-local effects	No	1

CHAPTER 3

COMSOL MULTIPHYSICS® – OVERVIEW AND CFD MODULE

OVERVIEW

In this chapter, we introduce COMSOL and its features as a software tool for modeling. The objective is to provide a "tour" of this software package, with emphasis on CFD module through using an example model, and introduce its main features, modules, and facilities as well as provide some guidelines for building models using COMSOL. To demonstrate COMSOL module applications, we will provide several modeling examples in detail in the next chapter. Because it would be exhaustive to include all features available in COMSOL in a single book, our main objective is to provide a collection of examples and modeling guidelines through which readers could build their own models. For those users who are new to COMSOL, we recommend to learn about, at least some of, the vast features of this tool by using its support resources, and consulting references like Tabatabaian [81]. In this overview we aim to cover COMSOL versions 4.4, and the recently published version 5. The main additions in version 5 relative to this book, are the Application

Builder tool and two new turbulence models (i.e. yPlus and L-VEL). For CFD modules, in version 5, tools for turbulent flow modeling through fans and grilles and automatic pipe connections to a 3D flow domain are available. Readers are referred to relevant COMSOL webinars and *http://www.comsol.com/release/5.0* for highlights of new features in version 5.

COMSOL is a flexible software multiphysics modeling tool, capable of modeling for example; structural mechanics, fluid flow, heat transfer, electromagnetic, and chemical reactions types of physics. It has many ready-to-use modules, for example CFD, yet it allows users to build their own models using its equation-solver facilities. COMSOL, which is a finite-element-based modeling tool, has a well-developed graphic user interface and several modules for modeling common and advanced types of physics involved in engineering and applied science fields. Recently the graphic user interface had a major upgrade in COMSOL 4.4, with some modification and add-on products in version 5, and with a new tool called Application Builder. The new interface makes it smoother, relative to previous versions, and 'guides' users to build their models through a Model Wizard feature. For highlights of version 4.4, please see *http://www.comsol.com/release/4.4*.

Building the geometry of a model is possible either by using CAD facilities (enhanced with a new product called Design Module, in version 5) available in COMSOL or by using live communication modules, such as LiveLink™. LiveLink is available for major CAD and computing packages (e.g. Inventor®, SolidWorks®, Excel®, Revit®, MATLAB®). In addition, users can import a solid model (part or assembly) in conventional format such as Parasolid, STEP, IGES, VRML, and STL. See the COMSOL manuals for CAD import modules, as version 5 allows users to import mesh and create model geometry from it.

Several Unit Systems of measurement are available in COMSOL. When a Unit System (e.g. *SI*) is chosen as default system, users can enter the data in different units but COMSOL automatically converts it to the main selected Unit System, for example *SI*.

Another major feature in COMSOL is the facility to solve any PDE/ODE that users might have and might not fit into classical governing equations (e.g. wave, heat, equilibrium) or relevant existing COM-SOL modules. This is done by using the Mathematics module. A new feature since version 4, called 0D allows users to solve problems that do not have space as a relevant defined dimension, such as electric circuits, chemical reactions, or thermal equivalent networks. Another recent feature, available in COMSOL is that users can run COMSOL directly through a CAD software package interface, such as SolidWorks® and some Autodesk® products. The author's experience with this package includes its ongoing improvement in modeling tools and features, especially the rich library of several equation solvers available, which makes it an efficient modeling tool for small-to-medium problems (in terms of geometrical size) especially when multiphysics modeling is involved. Meshing with COMSOL is almost seamless and 'automatic,' yet it provides users with the choice of having custom-designed mesh both for structured and unstructured types. Users can also import mesh files and create geometry from the imported mesh. In version 5, a new tool called Form Assembly is added, which can be used for meshing large assemblies of parts, and of having non-conforming meshes at the boundary. One of the features in Mesh facility, is that the intelligence exists based on the physics of the problem at hand. For example, if fluid flow modeling is involved then boundary layers elements are automatically generated close to solid walls, of course with some default settings. Users then, have the choice of modifying the default settings.

COMSOL has a rich materials property database and yet allows users to define their own database for their desired materials. In version 5, a new tool called Switch has been added, which can be used to solve a model for different materials by using the Sweep tool. In COMSOL, users have access and can see the governing equations related to the type of physics involved, explicitly through the graphic user interface on their computer screens; a feature that is very helpful and desirable, specifically for assigning right values to the variables and boundary conditions as well as in knowing what type of equations one is solving using the finite element method. COMSOL equations solver is capable of solving problems in stationary, time, or frequency domains, as well as solving a system of equations simultaneously or iteratively (segregated). The default solver, usually the best possible choice, is selected according to the physics of the problem at hand, yet users have the option of changing these settings according to their own choices. The post-processing features allow users to study, visualize, and build animation of their model results using color-coded surface graphs and data line graphs, as well as extracting the numerical results for further analysis. Finally, users can create a report document for the model problem at hand using the Reports facility. This feature enables users to generate a file using modeling results in common word processing formats.

For technical support, COMSOL has comprehensive and rich Help documentations as well as tutorials, available through its user interface. COMSOL provides free workshop and webinars for new users, as well as more extensive training courses available for purchase. In the following sections we will take a tour of COMSOL features and modules and introduce some of its features. More information about the features, models gallery, and tutorials is available on the COMSOL website (www.comsol.com).

APPLICATION BUILDER

A major upgrade in version 5 is the Application Builder tool. Application Builder provides excellent flexibility for communicating model results to other users so they can change some of the parameters/data in the model to evaluate and examine the relevant results. It is similar to having the built COMSOL model as an 'engine' in the background of the model application and using it 'implicitly' to run the model for different values of model parameters/data. The final result works like an 'App' which is commonly used with mobile media. Application Builder enables users to build and edit (currently in Windows® environment) a 'custom design' Graphic-User-Interface (GUI) for any model built in COMSOL Mutiphysics. The model Application, which can be password protected, can then run independently from (or jointly with) the original model. Through COMSOL Server™ (another new product that uses Amazon Cloud) users can run a model Application on their web browser, using a global floating license. Currently only model Application files can be uploaded to the COMSOL

Server. Following instructions can lead a user to the COMSOL Server for uploading/running a model application:

1. Sign in to COMSOL Server: www.comsol.com/try-comsol-server

2. Follow the instructions given, until you are logged in to the COMSOL Server page.

3. Run an application model from the Application library, or Upload a model Application and then run it.

Users can share their model Applications by clicking on the Administration tab, available on the COMSOL Server web page.

The Application Builder tool provides two Editors to build and edit the use-interface of the model application: (1) Form Editor, and (2) Method Editor. Users are referred to the COMSOL Application Builder Manual for further details.

We will use Application Builder and Form Editor for a model example given in this chapter. For further details see the COMSOL Application Builder Manual.

COMSOL MODULES

COMSOL has many ready-to-use modules to handle the modeling of most, if not all, commonly occurring engineering problems, for example Electrical, Mechanical, Fluids, Chemical related ones, including multiphysics couplings such as Joule heating, thermal stress, fluid-structure interaction, thermoelectric, and piezoelectric. In addition, users can solve unconventional governing equations/PDEs using available Mathematics modules in COMSOL.

Following is a list of COMSOL physics/application modules (i.e. add-on products) available for purchase or included in the main software platform. Additional features and modules are released with newer versions of the software. For an updated and complete list, check the COMSOL website (www.comsol.com).

CAD Import Module
Design Module

CFD Module
Pipe Flow Module
Structural Mechanics Module
Nonlinear Structural Materials Module
Fatigue Module
Multibody Dynamics Module
Heat Transfer Module
Optimization Module
AC/DC Module
Mathematics Module
Chemical Transport Module
Mixer Module
Microfluidics Module
Molecular Flow
Electrodeposition and Corrosion Modules
Acoustics Module
Batteries and Fuel Cells Module
Geomechanics Module
MEMS Module
RF Module
Wave Optics
Plasma and Semiconductor Modules
Subsurface Flow Module
Particle Tracing Module
Ray Optics Module

COMSOL MODEL/APPLICATION LIBRARIES AND TUTORIALS

After installing COMSOL, many other resources become available to the users to support their modeling tasks at hand. One of these resources is the Model Libraries, which offers solved models for training, teaching, or modification. Registered users can download these models and supporting documents (usually in PDF format). Models available in the libraries are useful in order to start a model with similar or closely-related physics and modify them according to a specific/desired modeling problem. A new Application library is available in version 5, which could be useful for applications built based on COMSOL models.

There are also two types of Help documents available for users under the Help button in the toolbar: Documentation and Dynamic Help. The Help Documentation offers users access to an extensive, searchable list of documents that explain interface icons and keys as well as details of modules, physics, meshing, geometry, post-processing, and more. Users at varying levels of expertise can refer to the documentation to find more details about COMSOL features as well as answers to their specific questions. The Dynamic Help feature opens a specific section of the Help documents relevant to the section or feature in use at hand. In version5, an Application library is added and gradually is populated with application model files. Similar to a COMSOL model, an Application model can be loaded and used for learning from its features and user interface layout.

COMSOL INTERFACE OR DESKTOP

After purchasing the product license you can install COMSOL on your machine, either PC or Mac. When launching COMSOL, the New window similar to the one shown in Figure 3.1 will open. COMSOL 5 has a new button, Application Wizard, for Application Builder. The default is for building a new model, either using the Model Wizard or a Blank Model. Users can also click on the File > Open, from the menu bar, and open an old file. It is recommended to start a new model using Model Wizard, since it will guide the users through the steps necessary for building a model.

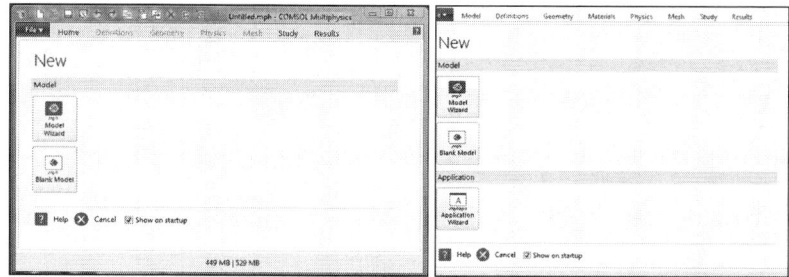

FIGURE 3.1: The New window opens when launching COMSOL 4.4 (left), or 5 (right).

After clicking on the Model Wizard icon in a new window, Select Space Dimension will open as shown in Figure 3.2. Users can choose the physical dimension of the model by clicking on the relevant icon, which includes 0D, 1D, and 2D Axisymmetric cases, as well. In order to proceed with this introductory 'tour' we select 2D Axisymmetric for our model example (laminar flow in a pipe with contraction, see Example 3.1) presented in this chapter.

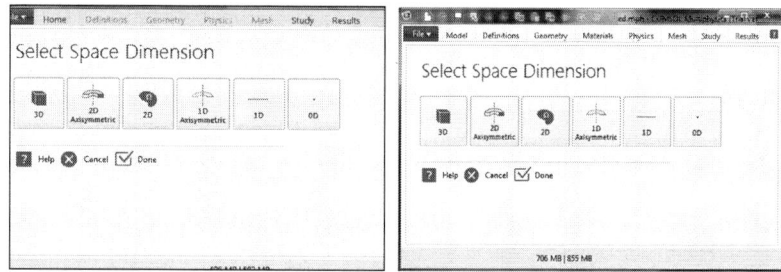

FIGURE 3.2: Select Space Dimension window in COMSOL 4.4 (left), or 5 (right).

After clicking on 2D Axisymmetric icon the Select Physics window opens as shown in Figure 3.3. Users can choose physics or multiphysics related to the problem at hand and assign it to the model. Available physics choices depend on the user's purchased license, including Modules.

For our example, we choose Fluid Flow > Single-Phase Flow > Laminar Flow from the list, as shown in Figure 3.3, and click Add button. For a selected choice, in this example Laminar flow, a brief explanation appears on the right side of the corresponding window. Next, we should select the type of solver for solving the model equations. In COMSOL, in general, solvers are selected based on the type of physics involved as default and labeled Study. Yet, users

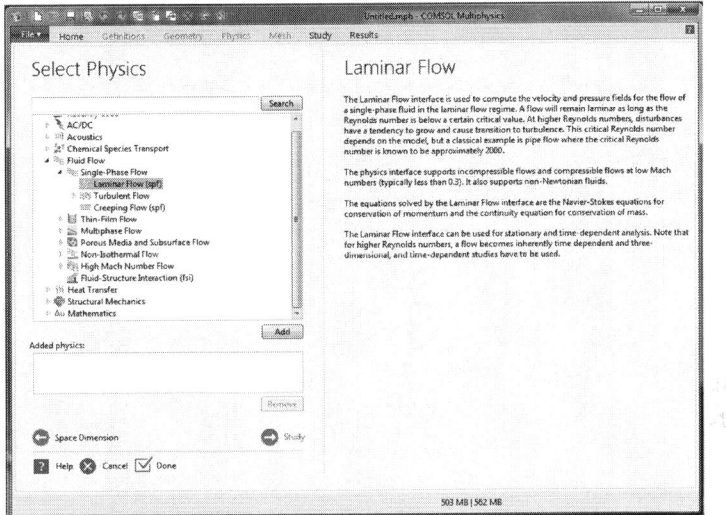

FIGURE 3.3: Select Physics window in COMSOL.

have the option to choose a different solver from the available list in COMSOL. This feature adds to the flexibility of COMSOL, as a comprehensive modeling tool. For this example click on the Study button, located at the bottom right corner of the Select Physics window. The Select Study window appears, as shown in Figure 3.4. Select Stationary by clicking on the relevant icon. At this stage, users have the option of going back to previous steps and modifying them

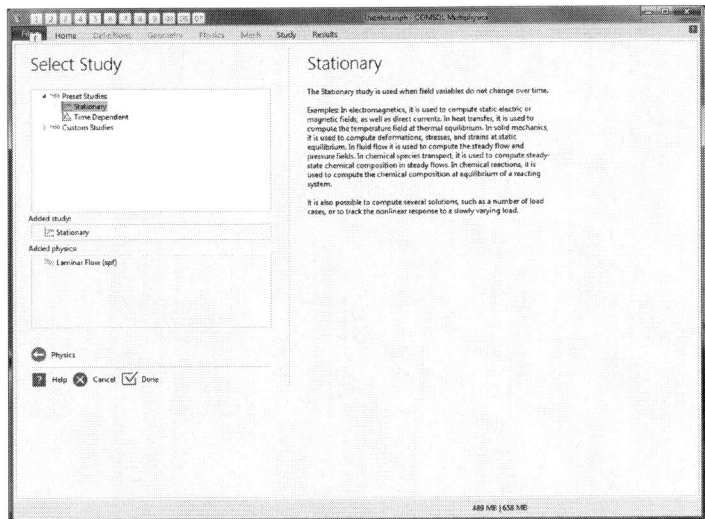

FIGURE 3.4: Select Study window in COMSOL.

if desired, accordingly. For this example, we keep the features that we have so far selected for this example and click on the Done button, located at the bottom right corner of the Select Study window.

The main interface or COMSOL desktop will appear. This interface includes a Quick Access Toolbar menu on the top, and a Ribbon bar which changes the corresponding ribbon Tabs selected. The Quick Access Toolbar could be moved to be placed under the Ribbon bar, as well. The Ribbon toolbar items are listed according to the, usual sequence used for building a model; Home (Model in version 5), Definitions, Geometry, Materials (in version 5) Physics, Mesh, Study, Results as shown in Figure 3.5. The Ribbon bar under Home/Model tab lists the modeling sequence actions required, as well.

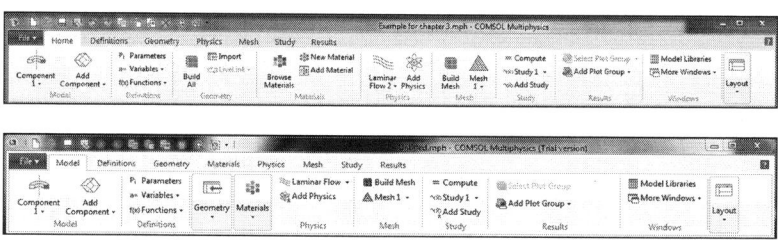

FIGURE 3.5: Toolbar and Ribbon for selected Home/Model tab in COMSOL 4.4 Desktop (top), and 5 (bottom).

A useful item in the Ribbon, under Home/Model tab, is the Layout. Users can choose their Desktop Layout by clicking on this icon, and choose for example, Reset Desktop. In addition, a COMSOL Desktop, as shown in Figure 3.6 for version 5, has three main sections or sub-windows; Model Builder, Settings, and Graphics which appear from left to right, respectively in the default Desktop layout. The Model builder or model tree window works as a registry for book keeping the model features, data, physics, mesh, study, results, etc. and can be used to quickly access model features or modify them if needed. The Settings window changes according to the item selected in the Model Builder. For example Geometry, as shown in Figure 3.6. Additional Information windows, like Messages, Progress, Log, Table, and External Process are also available, under More Windows tab from the Home/Model toolbar tab.

The third window is the Graphics window which shows the geometry of the model during building the model geometry, allows selection of domains and boundaries, and shows modeling results

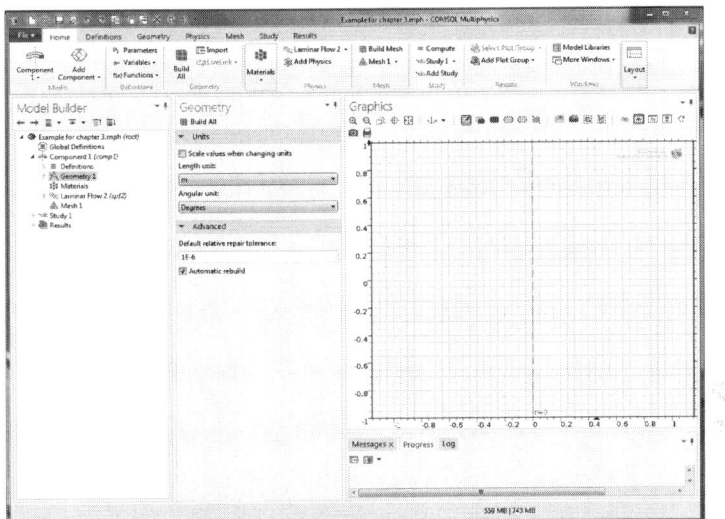

FIGURE 3.6: COMSOL 4.4 Desktop with Geometry data entry window.

including graphs. Any of these three Windows could be separated from the Desktop and moved around by making them Float. This feature is available by right-clicking on the corresponding window's title section and selecting Float option.

CFD MODULE

The COMSOL CFD module is available for purchase, in addition to the base software package for a fee. It contains ready-to-use modules for modeling a wide range of fluid flow problems, including single-phase flow, turbulent flow, non-isothermal flow, compressible flow, two-phase flow, flow in porous media, rotating machinery, thin-film flow, non-Newtonian flow, conjugate heat transfer, and reacting flow. The CFD module could be combined with other modules for multi-physics modeling, for example Fluid-Structure Interaction. Figure 3.7 shows a list of ready-to-use models available in CFD module of COMSOL. The list may differ based on modules installed on a computer.

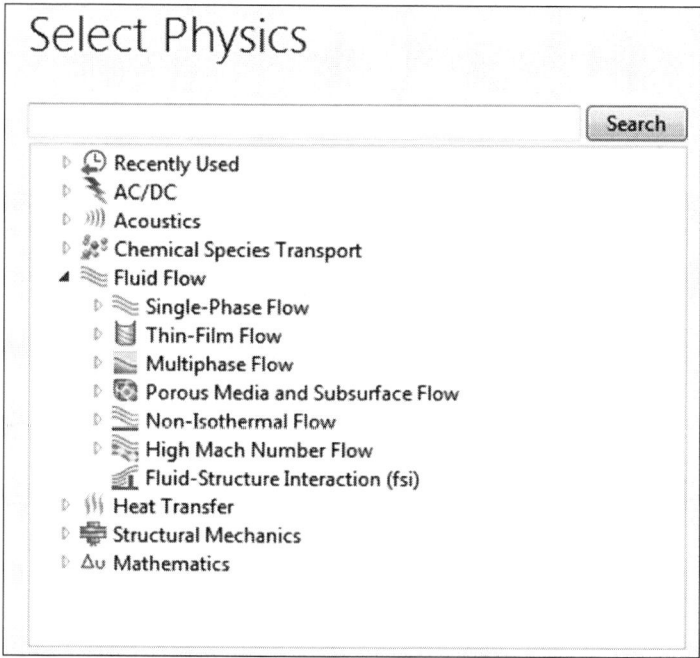

FIGURE 3.7: COMSOL CFD module ready-to-use models.

Each model category includes sub-models, for example Single-Phase Flow contains five turbulent flow models in COMSOL 4.4 or seven models in COMSOL 5, as shown in Figure 3.8. We will use these turbulent flow models in the next chapter (Chapter 4) and will provide worked-out examples to demonstrate their applications.

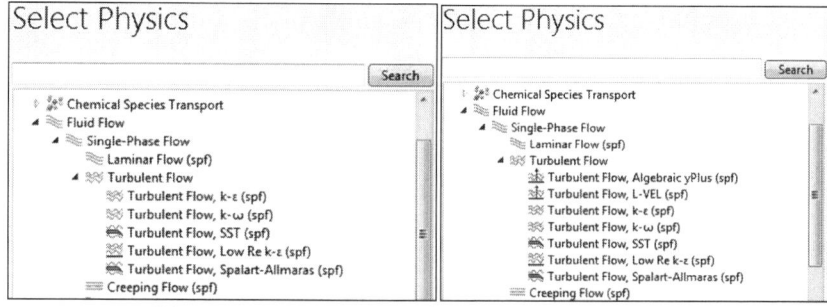

FIGURE 3.8: COMSOL 4.4 (left) and 5 (right) CFD module ready-to-use turbulent flow models.

Example 3.1: Laminar flow in a sudden pipe contraction

We continue with details of our example model to demonstrate, in general, the process and steps for building a model in COMSOL using the CFD module. Either versions, 4.4 or 5, could be used for building this example model, however version 5 provides the Application Builder tools. The problem considered here is flow of a fluid in a pipe with a sudden contraction. The flow is axisymmetric about the pipe axis, as shown in Figure 3.9, along with pipe sizes and lengths and coordinates system (r, z). The origin of the system of coordinates (r, z) is set to be at the centerline of the pipe with r-axis coinciding with pipe contraction location.

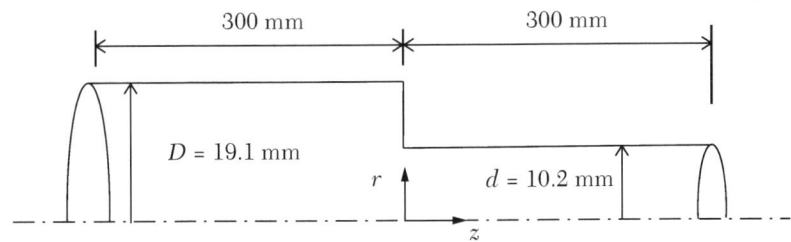

FIGURE 3.9: Geometry and dimensions for Example problem 3.1.

The boundary condition at the inlet of the pipe is considered as a fully-developed flow. Reynolds number based on the average fluid velocity U_{avg} and diameter D, related to the large-diameter pipe section is $Re = 372$. We calculate normalized velocity profile across the pipe, $w(r)/U_{avg}$ versus dimensionless distances along the centerline of the pipe, i.e. z/D (or z/d). We also validate the model results against experimental results of Durst et. al. [82].

Solution:

1. Launch COMSOL and click on the Model Wizard icon. From the Select Space Dimensions window options, click on the 2D Axisymmetric icon. The Select Physics window will open. Locate the Fluid Flow > Single-Phase Flow > Laminar Flow (spf) from the list and click on it and then click the Add button. The Laminar Flow (spf) will appear in the space under Added physics space. Then click on Study icon to move to Select Study window. In this window select Station-

ary and click on the Done button. The COMSOL Desktop window will appear. Save the file as Example 3.1.

2. Locate the Length unit selection in the Geometry/Settings window and change the unit to mm. For having all variables and dimensions, we use Parameters option, located in the Home/Model Ribbon tab option. Click on the Parameters icon and enter the data (case sensitive) as shown in Figure 3.10. Alternatively corresponding data could be loaded by clicking on Load from the File icon and open file params-example3.1.txt from the accompanying disc media.

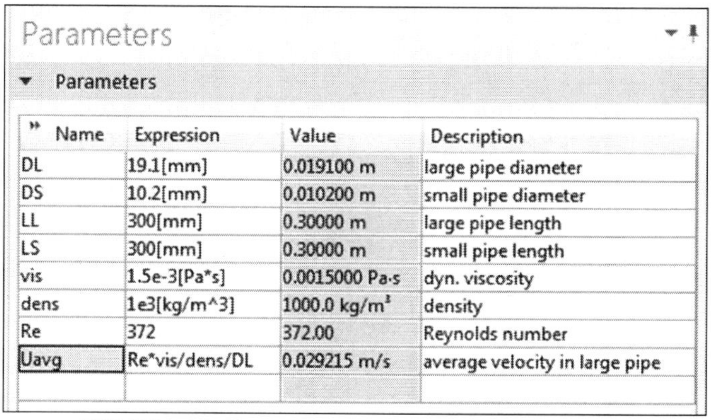

Parameters

▼ Parameters

»» Name	Expression	Value	Description
DL	19.1[mm]	0.019100 m	large pipe diameter
DS	10.2[mm]	0.010200 m	small pipe diameter
LL	300[mm]	0.30000 m	large pipe length
LS	300[mm]	0.30000 m	small pipe length
vis	1.5e-3[Pa*s]	0.0015000 Pa·s	dyn. viscosity
dens	1e3[kg/m^3]	1000.0 kg/m³	density
Re	372	372.00	Reynolds number
Uavg	Re*vis/dens/DL	0.029215 m/s	average velocity in large pipe

FIGURE 3.10: Flow parameters for Example problem 3.1.

3. Now we build the geometry of the pipe. Click on the Geometry tab in the Ribbon toolbar, select Rectangle, and move the cursor to the Graphics window and draw an arbitrary rectangle. The Rectangle 1 (r1) node will appears in the Model Builder window, under Geometry 1 node. Repeat the operation to create another rectangle, which will create Rectangle 2 (r2) node. (Alternatively, users can right-click on the Geometry 1 node in the Model Builder window and select Rectangle from the list). In the corresponding Rectangle/Settings windows enter the variables related to the dimensions of the pipe diameters with choosing Corner for the Base, as shown in Figure 3.11. Then, click on the Build All Objects button. Zoom in/out may be needed to see all the built pipe geometry in the Graphics window. To create the pipe contraction, click on the Graphics window and hit

Ctrl+A buttons, simultaneously on your keyboard. Then click on Difference (in version 5 this appears under the Booleans and Partitions tab) button in the Ribbon bar under Geometry Tab. The final 2D axisymmetric pipe geometry should be as shown in Figure 3.12. In this window the z-axis is in vertical direction, the r-axis is in horizontal direction, and the dash-dot line is the axisymmetric axis, located at $r = 0$.

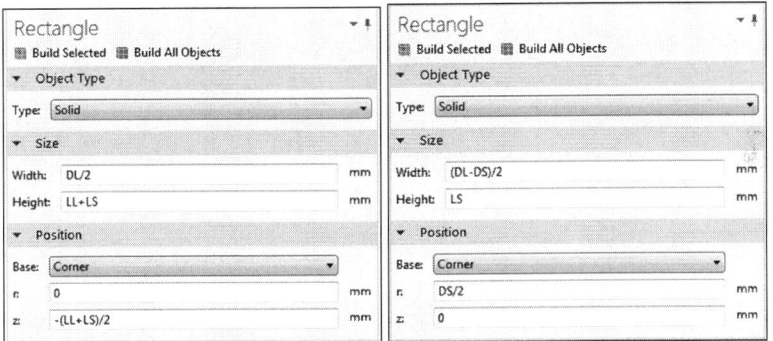

FIGURE 3.11: Dimensions for flow geometry for Example problem 3.1.

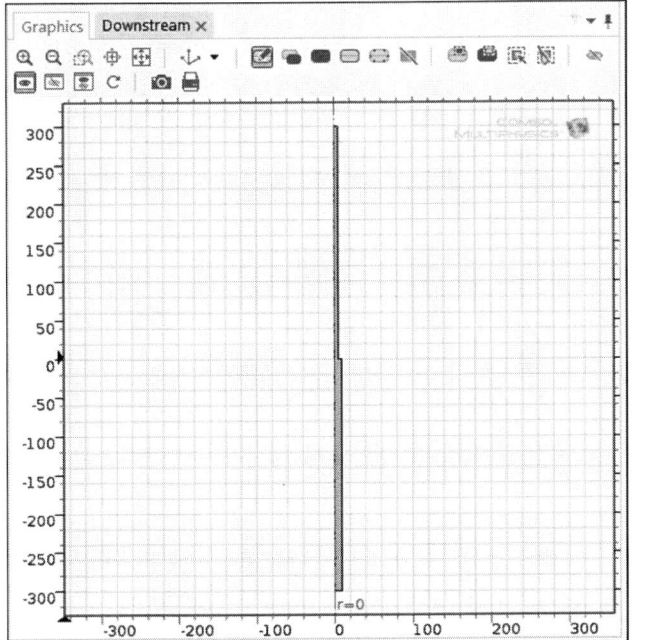

FIGURE 3.12: Pipe 2D-Axissymetric geometry for Example problem 3.1.

4. To define material properties of the fluid, click on the Fluid Properties 1 node in the Model Builder tree. In the Fluid Properties/Settings window locate the Fluid Properties section and select User defined option for both Density and Dynamic viscosity, and enter dens and vis, respectively, as shown in Figure 3.13. These entries should be the same as those defined in Parameters (i.e. they are case sensitive).

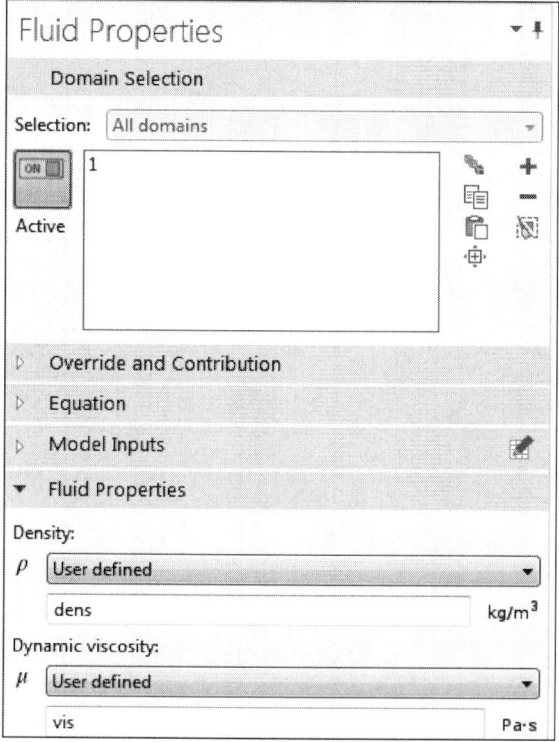

FIGURE 3.13: Fluid properties data for Example problem 3.1.

5. Now we define the boundary conditions. Click on the Physics tab in the Ribbon bar and from the list under Boundaries button select Inlet. In the Inlet/Settings window, add the inlet edge of the pipe (i.e. the lower edge of the pipe) to the Active list by moving the cursor to the Graphics window and clicking the corresponding edge located at $z = -300$. Boundary number 2 will appear in the list. Locate the Boundary condition section and select Laminar inflow from the list. Enter U_{avg} in the space below Average velocity, and

$5*\mathrm{Re}*\mathrm{DL}*0.06$ in the space under Entrance length. The Entrance length selection provides a fully developed flow at the entrance of the pipe. Finally, in the same Inlet window, check the box for Constrain endpoints to zero. Similarly, create an Outlet boundary with zero pressure, i.e. default value and assign pipe exit or boundary edge number 3 located at $z = 300$ to it, as shown in Figure 3.14. The remaining boundaries are no-slip wall type, selected by default. Users can check this by clicking on the Wall1 node in the Model Builder tree.

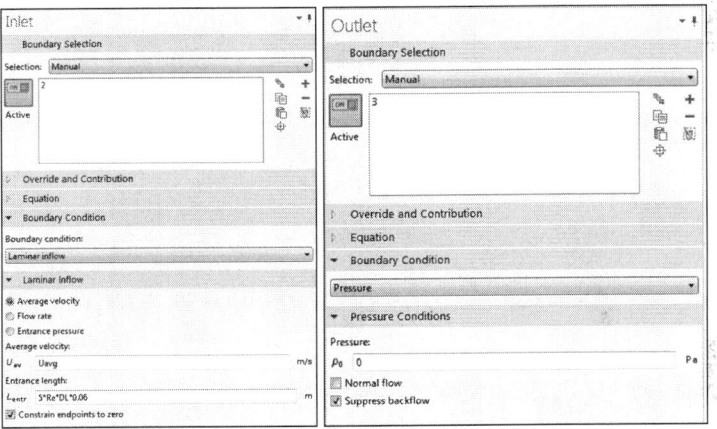

FIGURE 3.14: Boundary conditions for Example problem 3.1.

6. With the geometry and boundary conditions set up, we will now build a mesh. Click on the Mesh 1 node in the Model Builder window. From the Mesh/Settings window locate Element size section and select Coarse from the list and click the Build All button. The total number of elements, including boundary layer type elements, appears in the Message window. Note the default mesh Sequence type, i.e. Physics-controlled mesh. This type is selected by default according to the physics set for the model. In this example a mesh with dense element distribution is built close to the pipe walls suitable for modeling the boundary layer region with high velocity gradient. A close-up of the elements close to the pipe contraction is shown in Figure 3.15.

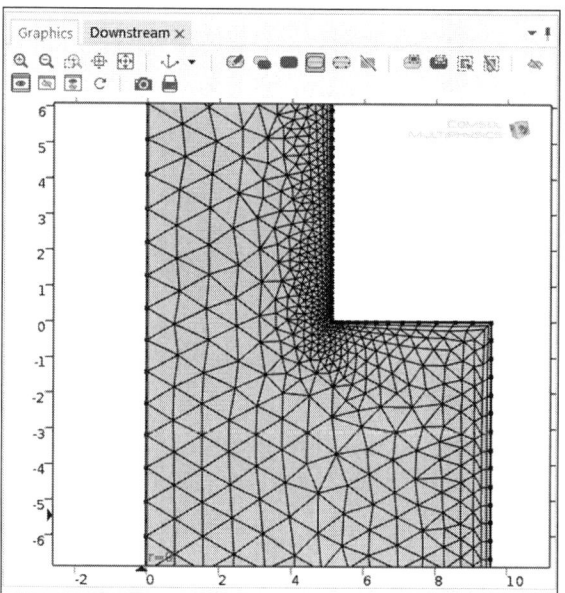

FIGURE 3.15: Boundary and domain hybrid mesh close to pipe contraction for Example problem 3.1.

7. Now we start the computation by running the model. Click on the Home/Model tab in the Ribbon bar and select Compute. Default results for Velocity, Pressure contour, and Velocity 3D will appear, after computation is finished. These results are shown in the Graphics window, by clicking on corresponding nodes located under Results in the Model Builder tree. Figure 3.16 show typical results close to the pipe contraction. Objects can be moved and zoomed in/out, etc., using tools located in the Graphics window's toolbar.

8. In order to compare the results with experimental ones and demonstrate post-processing Line Graph tools, we manipulate the obtained results by calculating the dimensionless fluid velocity profiles at several cross-sections along the pipe axis. Click on the Results tab from the Ribbon tool bar and select 1D Plot Group. A new node (1D Plot Group 4) will appear under the Results node, in the Model Builder window. Rename this node Upstream graphs. Click on the Line Graph button from the list under the Upstream graphs tab, in the Ribbon tool bar. In the Line Graph/Settings window,

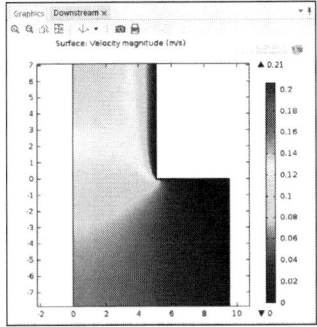
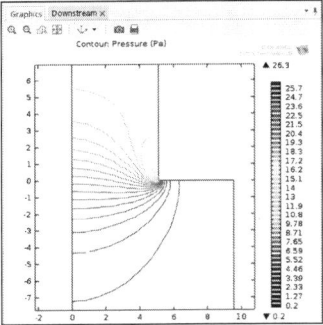

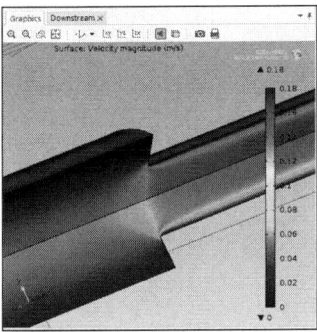

FIGURE 3.16: Model results for velocity and pressure contour close to the pipe contraction region.

locate the y-axis Data section and type $r/(DL/2)$ in the space provide under Expression. In the same window locate the x-Axis Data section and select Expression from the list under Parameters. Enter w/U_{avg} in the space under Expression. Rename the Line Graph1 node velocity at $z/D = -0.523$.

9. To extract data from the model Solution at this location along the axis of the pipe, we should define a cross-section or a Cut Line 2D. Expand the list under the Results node and right-click on Data Sets and select Cut Line 2D. A new node (Cut Line 2D 1) will appear in the Model Builder window. Rename it Data at $z/D = -0.523$, click on it, and in the Cut Line 2D/ Settings window locate the Line Data section, select Point and direction from the list (in front of Line entry method) and enter 0 for r, and $-0.523*DL$ for z. This will extract numeri-cal values for a cross-section at $z - 0.523 \times DL = 9.9893$ mm upstream from the pipe contraction. Now we assign this data

set to the line graph that we have created. Click on velocity at $z/D = -0.523$ node under Upstream graphs and in the corresponding Line Graph window locate the Data section and select Data at $z/D = -0.523$. To change the line style and create markers, Expand Coloring and Style section and enter 2 for Width and select Plus sign for marker from the Marker selection. All data and set ups are ready for this graph; click Plot. The result line graph is shown in Figure 3.17. Graph and Axes titles and scales can be changed using the tools available in 1D Plot Group and Line Graph windows.

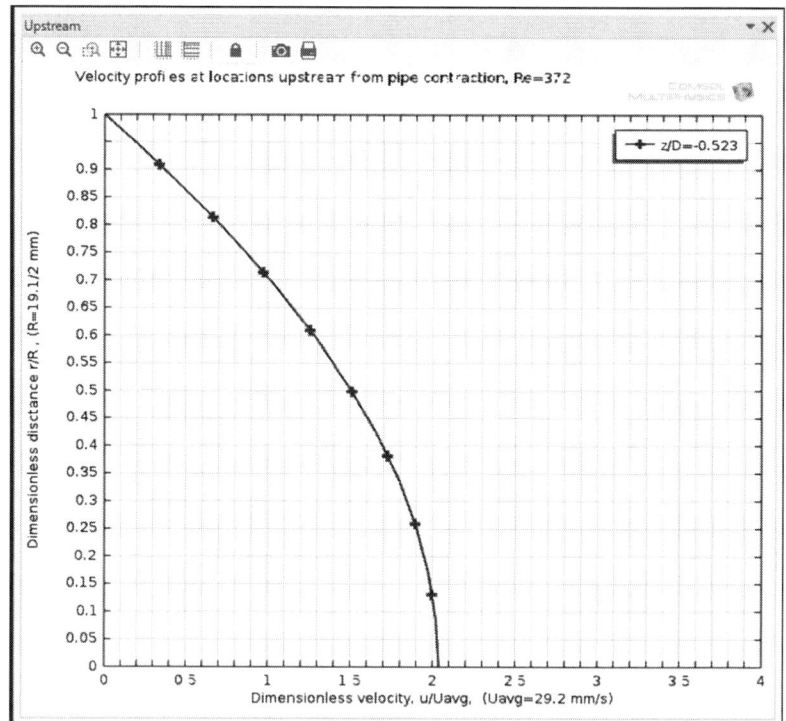

FIGURE 3.17: Model results for velocity across the pipe at upstream from the pipe contraction.

10. Repeat instructions given in Steps 8–9, to create velocity profiles at five more locations along the pipe axis, according to Figure 3.18. Final graphs are shown in Figure 3.19. Graph title and axis scales are modified according to the range of variables involved.

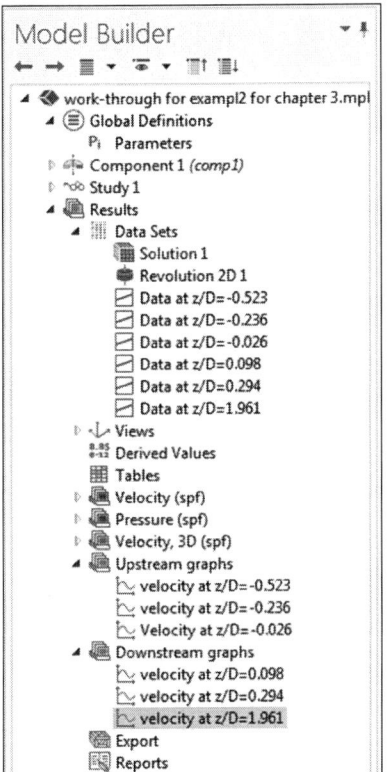

FIGURE 3.18: Model line graphs data set up for velocity profiles.

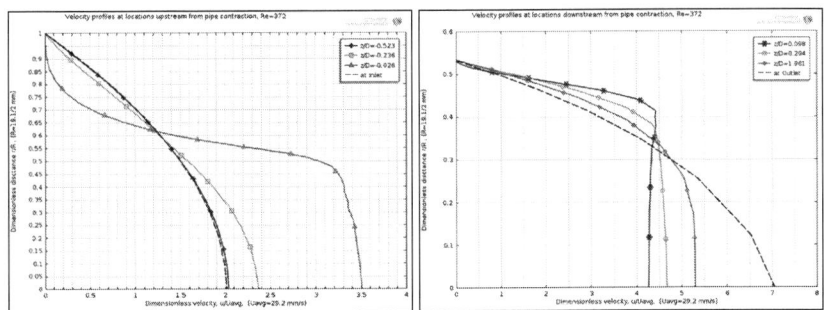

FIGURE 3.19: Model results for velocity profiles across several locations along the pipe axis.

Finally, we compare the model results, for dimensionless axial velocity, against existing experimental ones [82], as shown in Table 3.1 (numerical values may show small variations depending on calculation precisions and/or mesh resolution used). In order to extract numerical values from the model solution we should create the physical points in the geometry and evaluate the numerical values of the desired variables, for example for velocity, at these points.

TABLE 3.1: Comparison of model against existing experimental results [82], for dimensionless axial fluid velocity.

z/D	U/Uavg					
	−0.523		−0.236		−0.026	
2r/D	Model	Exp.	Model	Exp.	Model	Exp.
0.08	2.02612	2.091	2.33699	2.410	3.46273	3.409
0.24	1.91191	1.977	2.16949	2.245	3.37023	3.306
0.52	1.46458	1.524	1.51	1.565	1.96339	1.864
0.76	0.83488	0.824	0.7359	0.762	0.10423	0.062
0.98	0.07559	0.051	0.04821	0.051	−6.47E-04	−0.010

11. Click on the Geometry tab, and from the list under More Primitives select Point. Then click anywhere in the flow domain. A new node, Point 1(*pt1*), will appear under Geometry 1 in the Model Builder tree. Click on the Point 1 (*pt1*) node and in the Point/Settings window enter 0.08∗*DL*/2 and −0.523∗*DL* for *r* and *z*, respectively. Click the Build Selected button; a physical node will appear (may require zoom in/out and a click on Geometry 1 node) in the flow domain at and , which is exactly what we want to have, according to Table 3.1. Click the Compute button (located in Home tab selections). To evaluate the value of axial velocity at this Point, right-click on the Derived Values node, located under Results, and select Point Evaluation from the list. In the corresponding window, add Point 1 to the Selection list, by clicking on it, in the Graphics window. Locate the Expression section and enter w/U_{avg}. Click the Evaluate button, located on the top of Point Evaluation window. The result will appear in the Table 1 window as 2.02612. This numerical value is compared against the corresponding experimental value, 2.091. amd measured at the same point [82]. Results for numerical values may differ based on the mesh resolution and, in general could be improved to match closer to the experimental results.

Another tool, which could be used, is the tool Results >Data Sets >Cut Point 2D. This tool extracts data at desired physical points in the flow domain from the solution and in conjunction with the tool Results >Derived Values >Point Evaluation we can calculate the result. This way, we don't need to run the model after each node is created in the geometry. Therefore we save several runs, but we don't see the physical points in the Graphics window (as shown in Figure 3.20, for V4.4).

12. Repeat instructions given in Step 11, to create 14 more points and evaluate their corresponding fluid axial velocities. Renaming the nodes created in the Model Builder tree would be useful and are recommended. The final result is shown in Figure 3.20.

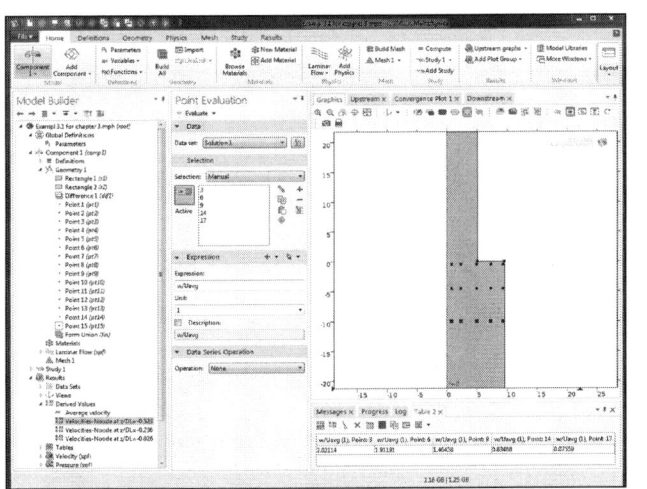

FIGURE 3.20: Set up for evaluation of numerical values of axial velocity at several locations along the pipe axis.

This concludes this example; users may enhance the accuracy of the results by examining different mesh resolutions and types, as well as evaluating the involved variables at cross-sectional locations downstream of the pipe contraction.

USING APPLICATION BUILDER FOR EXAMPLE 3.1

In this section, we use this tool for creating an Application for the model Example 3.1. Users who are using COMSOL4.4 should upgrade to version 5, in order to have the Application Builder facilities available to them. However if a model is built in version 4.4, it can be uploaded/opened in version 5, however the uploaded file (once saved) can't be opened again in version 4.4 (i.e. COMSOL model files are not retroactive).

Solution:

1. Launch COMSOL 5 and click on the Application Wizard icon. In the Select Model for Application window, click on the Browse icon and locate/select that which was saved for Example 3.1. The New Form window appears, as shown in Figure 3.21. Save the file as App-Example 3.1.

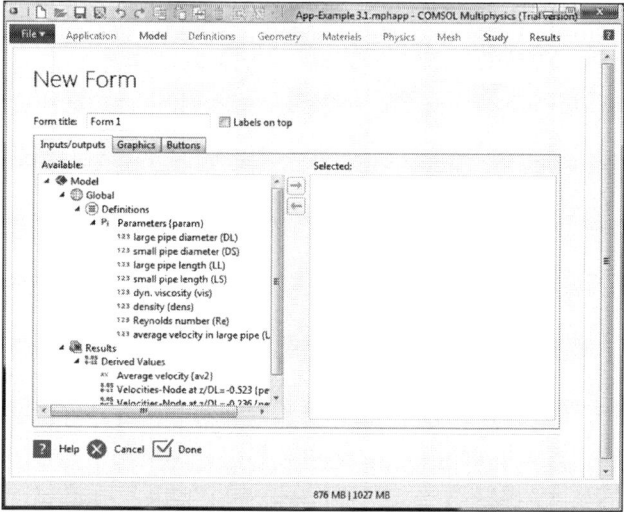

FIGURE 3.21: Application Builder New Form window for Example 3.1.

2. Change the Form title to Laminar flow in a sudden pipe contraction. Check the box for Label on the top. Under the Inputs/outputs tab, double click on the Reynolds number (Re) node in the Available list, in order to move it to the Selected list. Also, under Results, double click on Velocities-Node at $z/DL = -0.523$, Velocities-Node at $z/DL = -0.236$, and Velocities-Node at $z/DL = -0.026$. The result is shown in Figure 3.22.

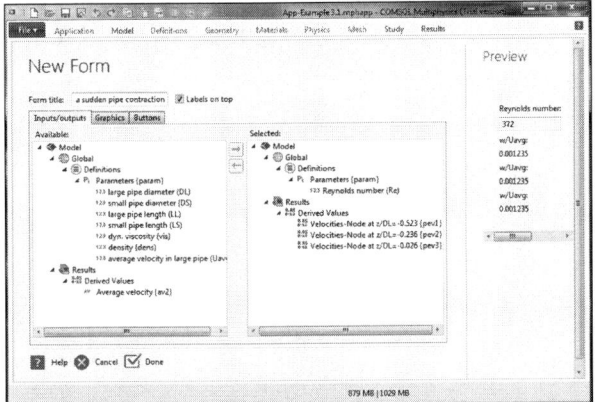

FIGURE 3.22: Application Builder Selected Input/outputs for Example 3.1.

3. Click on the Graphics tab, in the New Form window. From the Available list, double click on Velocity 3D (spf), Upstream graphs, and Downstream graphs, to move them to the Selected list.

4. Click on the Buttons tab, in the New Form window. From the Available list, double click on Compute Study 1, in order to move it to the Selected list. Click on the Done icon. The Form Editor Desktop (or Application Builder Desktop) appears, as shown in Figure 3.23.

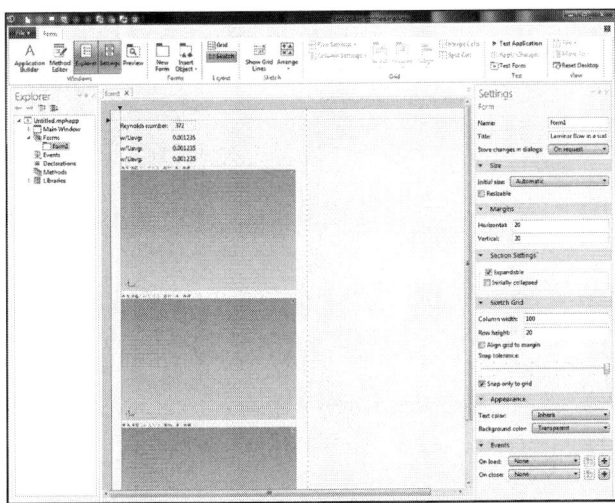

FIGURE 3.23: Application Builder Desktop window for Example 3.1.

5. Rearrange/resize the Graphics windows, to a desired layout format by drag-and-drop operations. Also move the Compute button to the top. Now we assign a picture to the Compute button. Click on the Compute button and from the Settings window locate Picture and select compute_32.png and from Size select large. Occasionally click on the Test Form button in the ribbon toolbar to see the result layout. Now click on the Test Application button. The App window will appear, as shown in Figure 3.24. In this window, change the Reynolds number to 196 and click on the Compute button. Watch the graphs and outputs that will change after computation is done. Users can use the toolbars associated with each graphic window, such as Zoom in/out, etc., to manipulate the graphs. Save the file. As mentioned in the previous section, users may share this model Application example by uploading it to the COMSOL Server.

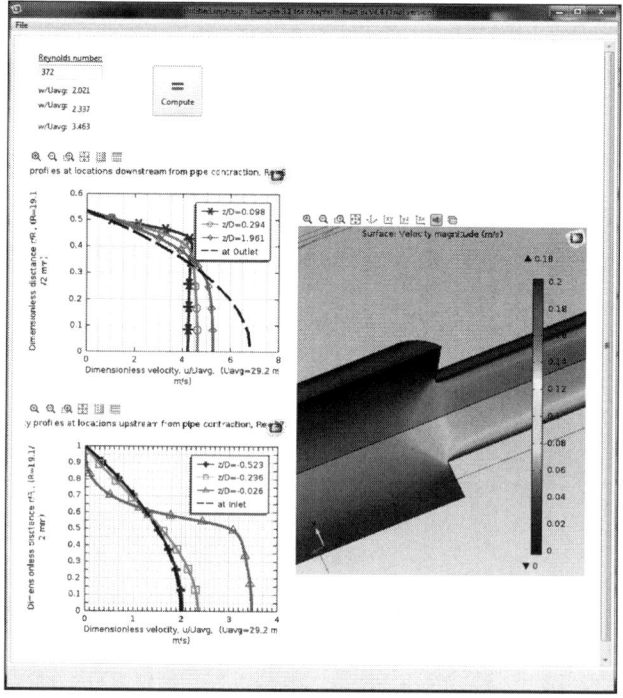

FIGURE 3.24: Application for Example 3.1, built by COMSOL 5 Application Builder tool.

GENERAL GUIDELINES FOR BUILDING A MODEL IN COMSOL

In general, major sequential steps for building a model using COMSOL for a given problem are as follows:

1. Define the problem, including physics and materials involved.

2. Identify the governing equations and boundary conditions to have a clear understanding of the scope of the problem's solution.

3. Launch COMSOL.

4. Use COMSOL features to assign the dimension (1D, 2D, 3D, etc.), the physics involved, and the temporal (steady or transient, etc.) of the problem.

5. Build the geometry of the problem (if required), import your CAD file, or use LiveLink to access your model geometry.

6. Assign material properties to the built geometry blocks of the problem.

7. Add physics and boundary conditions according to steps 1 and 2.

8. Create a mesh or finite elements for the built geometry.

9. Solve/run the model and verify the results.

10. Visualize the results and validate them, either using hand calculations or comparing to known results, either analytical, experimental, or validated numerical ones.

11. Create a model Application, available in COMSOL 5.

12. Create a report for the model that includes its specifications.

EXERCISES

3.1. Using the Instructions given in Chapter 3, build the model example given for Example 3.1.

3.2. Using the model solution for Example 3.1, create two more meshes with higher resolutions for this model and investigate the mesh independency of the results, as well as validating them against experimental results given in Table 3.1.

3.3. For Example 3.1, create points across the pipe at two locations upstream the pipe contraction and draw the fluid axial velocity profiles at these locations.

3.4. Modify Example 3.1, with adding another pipe section downstream to its existing small-diameter pipe and model the new problem.

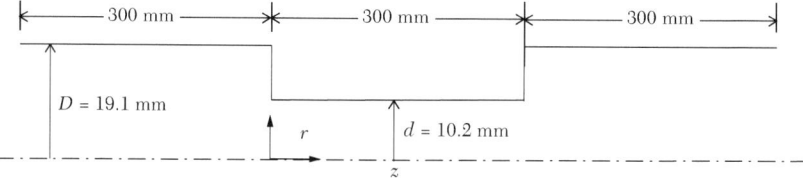

3.5. Modify Example 3.1, by adding another large–diameter pipe section upstream to its existing large-diameter pipe, and model the new problem.

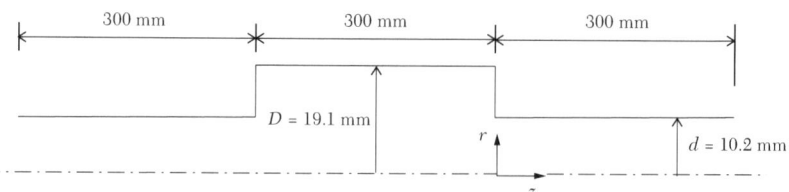

3.6. Using the Application Builder tool (available in COMSOL 5), build an application for Example 3.1. Try several layouts using Form Editor Desktop tool.

4

TURBULENT FLOW MODELS– APPLICATIONS

OVERVIEW

In this chapter, we present applications of turbulent models available in COMSOL through worked-out examples. Different types of flows including; internal flow, flow around objects and airfoils, free flow, and flow with separation will be modeled. The emphasis will be on industrial flows and examples with potential practical applications. For the purpose of comparison and validation, the obtained modeling results of the examples provided are compared against existing experimental data or numerical solutions, where and when available. Users might make use of several resources available, mainly those from NPARC (National Program for Applications-Oriented Research in CFD) Alliance of NASA. NPARC archive [83] offers a comprehensive list of cases for different flows. In order to define V&V (i.e. verification and validation), we refer to "AIAA Guidelines" [84]. In this Guideline verification is defined as: *The process of determining that a model implementation accurately represents the developer's conceptual description of the model and the solution to the model,* and Validation is defined as: *The process of determining the degree to which a model is an accurate representation of the real world from the perspective of the intended uses of the model.*

We recommend that users familiarize themselves with the COMSOL interface or Desktop and tools available with it, before trying the examples provided in this chapter. COMSOL provides webinars which could be used for this purpose and are available online.

Example 4.1: Modeling of turbulent flow for an asymmetric diffuser

For this example we model the turbulent flow in an asymmetric planar diffuser which has a reliable experimental database, after Buice [85]. Fluid flow from the narrow section of a diffuser enters into the wider section and hence flow separation happens due to adverse pressure gradient which in turn results from flow deceleration. The overall increase in drag force, in addition to the size of the separation, pressure variation, and energy losses are of interest for industrial applications. Capturing the detachment and reattachment point of the separation region is challenging, and a suitable turbulence model which can capture the flow in the diffuser zone should be selected. To capture the flow separation we have the choices of the k-ω model, SST model, or S-A model (see Chapter 2). We choose the SST model for this example. However, it is recommended that users try k-ω model and/or S-A model as well, for comparison [86]. Recall that SST model (i.e. the version available in COMSOL) does not require a wall function whereas the k-ω model requires application of a wall function.

A schematic of the diffuser geometry considered for this example is shown in Figure 4.1, with H = 1.5 cm, [85]. Reynolds number, based on upstream data is 20,000.

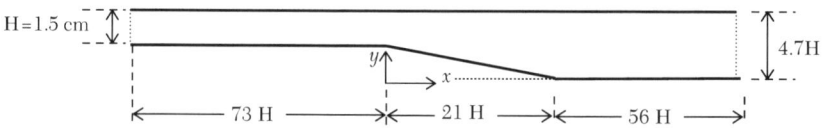

FIGURE 4.1: Schematic geometry of planar asymmetric diffuser.

Solution:

1. Launch COMSOL and from File>Save as, in the New window, save the model as Example 4.1. Click on the Model Wizard icon.

2. From the Select Space Dimension window, click on the 2D icon. The Select Physics window will appear. Select Fluid Flow>Single-Phase Flow>Turbulent Flow>Turbulent Flow, SST (spf). Then click on the Add icon. Figure 4.2 shows the Select Physics window and options selected. Note the list of dependent variables.

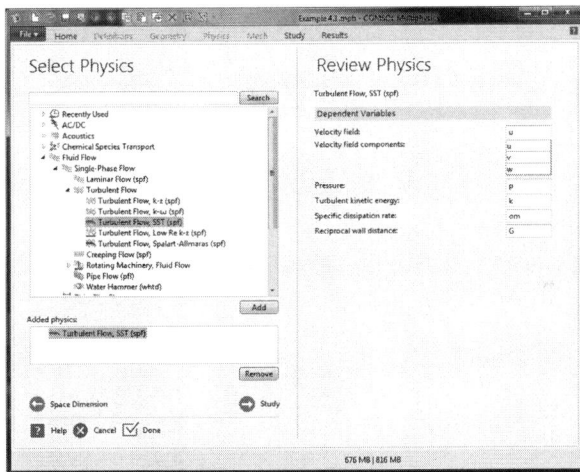

FIGURE 4.2: Select Physics window and options for Example 4.1.

3. Click on the Study icon/arrow and, from the Select Study window, under Preset Studies, click on Stationary with Initialization. Click on the Done icon. The COMSOL Desktop interface appears, as shown in Figure 4.3. In the Geometry window, locate Units section and change the Length unit to cm.

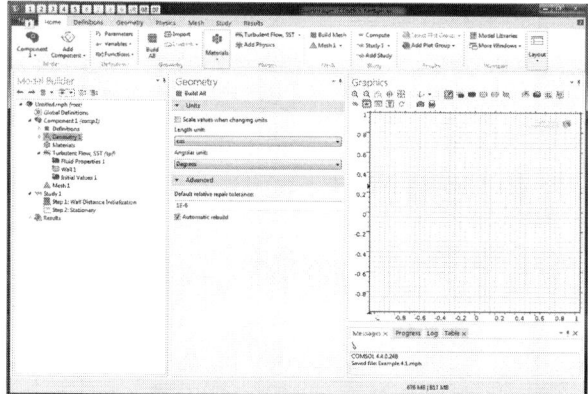

FIGURE 4.3: COMSOL Desktop interface for Example 4.1.

4. Now we make a list of input data used for this model as
parameters. From the Home /Model tab ribbon, click on
Parameters. In the Parameters window, enter the data (case
sensitive) as shown in Figure 4.4. Alternatively this data can
be imported from the accompanying disk.

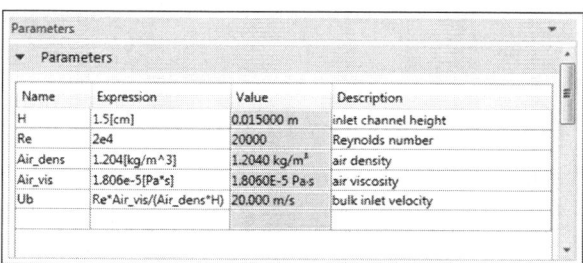

FIGURE 4.4: Parameters data for Example 4.1.

5. To build the geometry of the diffuser, click on the Geometry tab in the Ribbon and select the Draw Line tool from
the list. Then move the cursor to the Graphics window and
draw a straight horizontal line. To release the cursor, right-
click. In the Model Builder window, expand the Geometry
1 node and click on the Bezier Polygon 1 (b1) node. In the
corresponding Bezier Polygon window, locate Polygon Segments and click on Segment 1 (linear). Enter the data in the
Control points, according to Figure 4.5, and click on Build
Selected. You may have to zoom in for the line to appear in
the Graphics window. To add more lines, click on the Add

Linear button four times, and enter the data for Segment 2(linear) to Segment 5(linear), as shown in Table 4.1. Click on Build All Objects. The final Geometry of the diffuser is shown in Figure 4.6, as it appears in the Graphics window.

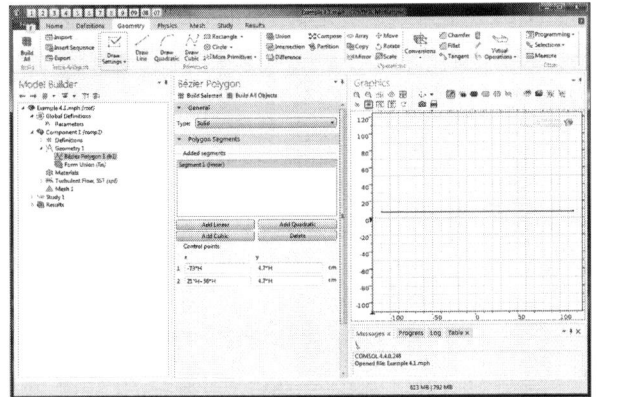

FIGURE 4.5: Geometry data and interface window for Example 4.1.

TABLE 4.1: Coordinates for line segments of diffuser geometry.

Line	Control points		
Segment 1 (linear)	x	y	
	1 -73*H	4.7*H	cm
	2 21*H+56*H	4.7*H	cm
Segment 2 (linear)	x	y	
	1 21*H+56*H	4.7*H	cm
	2 21*H+56*H	0	cm
Segment 3 (linear)	x	y	
	1 21*H+56*H	0	cm
	2 21*H	0	cm
Segment 4 (linear)	x	y	
	1 21*H	0	cm
	2 0	4.7*H-H	cm
Segment 5 (linear)	x	y	
	1 0	4.7*H-H	cm
	2 -73*H	4.7*H-H	cm

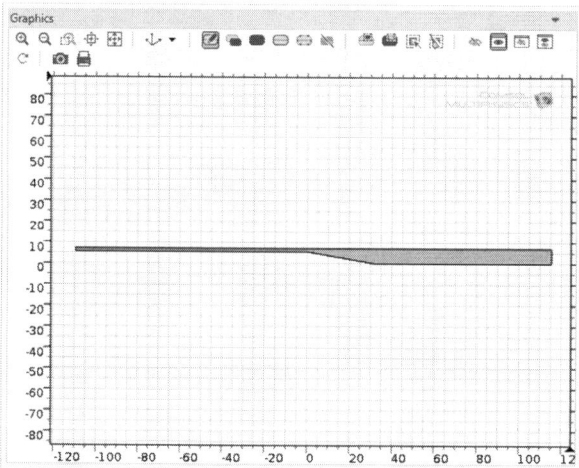

FIGURE 4.6: Geometry of diffuser shown in Graphics window.

6. To assign fluid properties to the flow domain, click on the Turbulent Flow, SST (spf)>Fluid Properties 1 node in the Model Builder window. In the corresponding window, locate section Fluid Properties and select User defined from Density list and enter Air_dens. Similarly for Dynamic viscosity, enter Air_vis. See Figure 4.7. The Reference length scale: l_{ref}, is left as default [55], which gives a value of one-tenth of the smallest diffuser length scale, i.e. 1.5 mm.

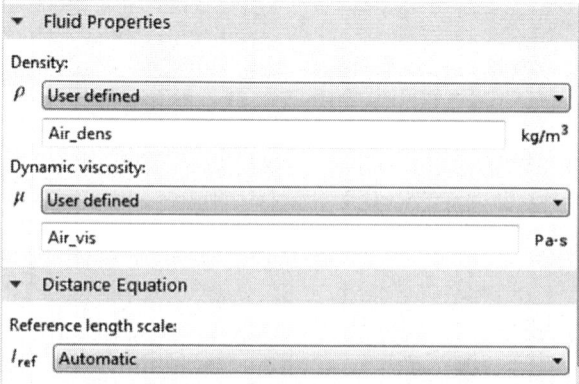

FIGURE 4.7: Fluid properties data entry section.

7. For boundary conditions at the inlet, click on the Physics tab and select Input from the list under Boundaries. In the corresponding window for Inlet, assign boundary 1 by clicking on the corresponding edge of the diffuser geometry. Enter 0.05 for Turbulence intensity and 0.07*3*H/2, (see Table 2.2), for Turbulence length scale and Ub for Velocity, as shown in Figure 4.8. Similarly, create the Outlet and assign boundary 6 to it. We leave pressure at default value of zero.

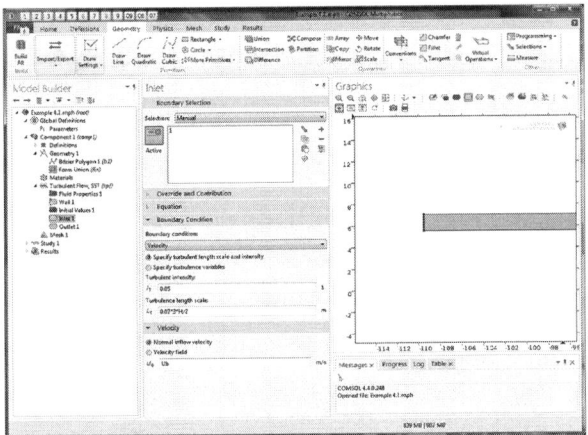

FIGURE 4.8: Inlet data entry section and boundary assigned.

So far we have built geometry, physics, and boundary conditions. The model is ready for building the mesh.

8. Click on the Mesh 1 node in the Model Builder window. In the Mesh window, leave the default for Sequence type as Physics-controlled mesh, and for Element size as Normal. Users may want to create a coarser or Mapped mesh first and use the corresponding model result as initial solution for a relatively finer mesh. Click on Build All to build a total of 175,736 hybrid elements, consisting of 5,020 boundary layer elements. A zoom-in of the mesh at the diffuser ramp is shown in Figure 4.9.

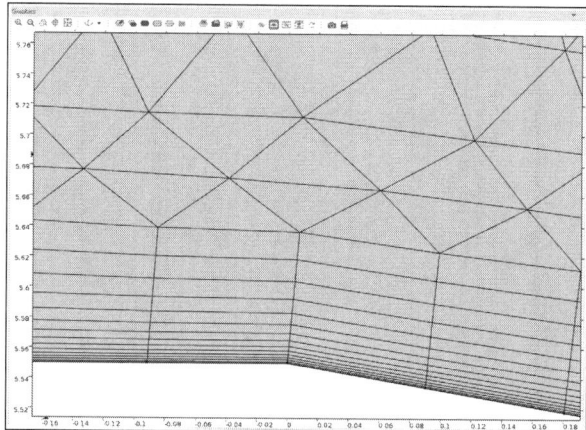

FIGURE 4.9: Mesh built consisting of boundary-layer and triangular elements.

9. The model is ready for computation. Right-click on the Study 1 node, in the Model Builder and click on Compute. Users can open the Convergence Plot 1 window to see the progress of the solution convergence per iteration numbers. On a typical computer, it takes close to one hour for computation to finish. High demand on computer power and time is one of the 'characteristics' of the SST model. To minimize these requirements, users may want to run the model using the k-ε model and then use the result as the initial solution for SST model. This approach also helps the convergence. A user-defined mesh could also help with these constraints.

10. Default results for mean velocity magnitude appear in the Graphics window, along with corresponding nodes for pressure and wall resolution (dimensionless distance to cell center) that appear in the Model Builder window. A zoom-in set of results for velocity and pressure contours are shown in Figure 4.10.

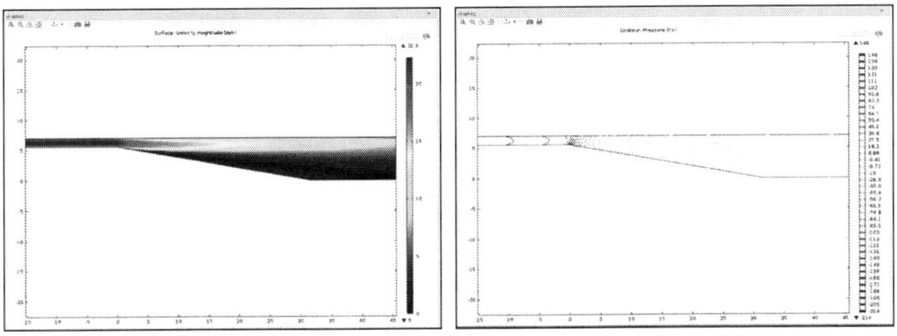

FIGURE 4.10: Mean velocity and pressure contours along diffuser.

In order to compare our results with the existing experimental results [85] and [83], we create line graphs at certain cross-sectional locations of the diffuser. We plot the axial velocity profiles at these locations for the purpose of comparison with the experimental results. The selected locations are at

$$\frac{x}{H} = -5.82, 2.5558, 5.9442, 13.468, 16.836,$$

$$26.227, 33.931, 66.94.$$

11. To extract data for desired cross sections, expand the Results > Data Sets node, in the model tree, and right-click on Data Sets to select Cut Line 2D. The Cut Line 2D 1 node appears. In the corresponding settings window, locate the Line Data section and choose Point and direction from the Line entry method list. Enter the data as shown in Figure 4.11. Similarly, generate seven more cross-sections by creating Cut Line 2D 2 to 8, and by using data given for selected locations (i.e.,

$$\frac{x}{H} = 2.5558, 5.9442, 13.468, 16.836, 26.227, 33.931, 66.94)$$

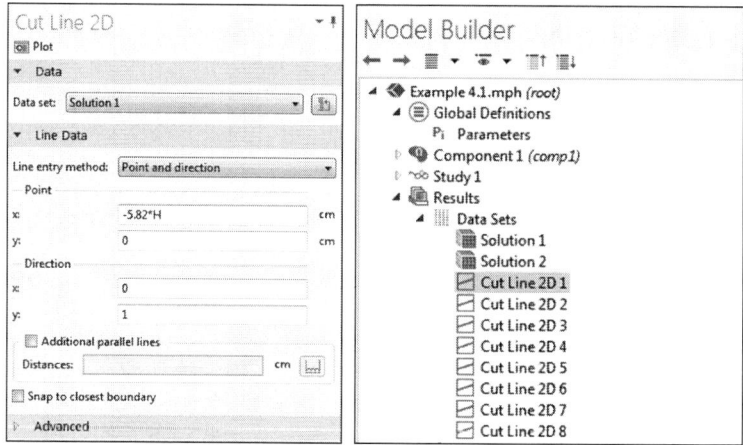

FIGURE 4.11: Window for cross-sectional data entry, data for Cut Line 2D 1 is shown.

12. Now having the data for each cross-section, build line graphs for each one. Click on the Results tab in the toolbar, and select the 1D Plot Group. From the list, select Line Graph. In the Line Graph window, locate the Data section and select Cut Line 2D 1 from the list. Enter the data according to Figure 4.12 for y-Axis Data and x-Axis Data sections. The Coloring and Style section is optional for manipulation of the appearance of the graphs. Click on the Plot icon, located on the top of window. Similarly, create seven more Line Graph 2–8. Note that for each graph corresponding cross-sectional data should be used (i.e. for Line Graph 2 use Cut Line 2D 2, etc.). Results for dimensionless mean axial velocity u/Ub are shown at desired cross sections. For demonstration and comparison purposes, a factor of 10 and location distance of x/H is used. Figure 4.13 shows a comparison of the model results against those of Study #1 of NPARC [83], which are in good agreement.

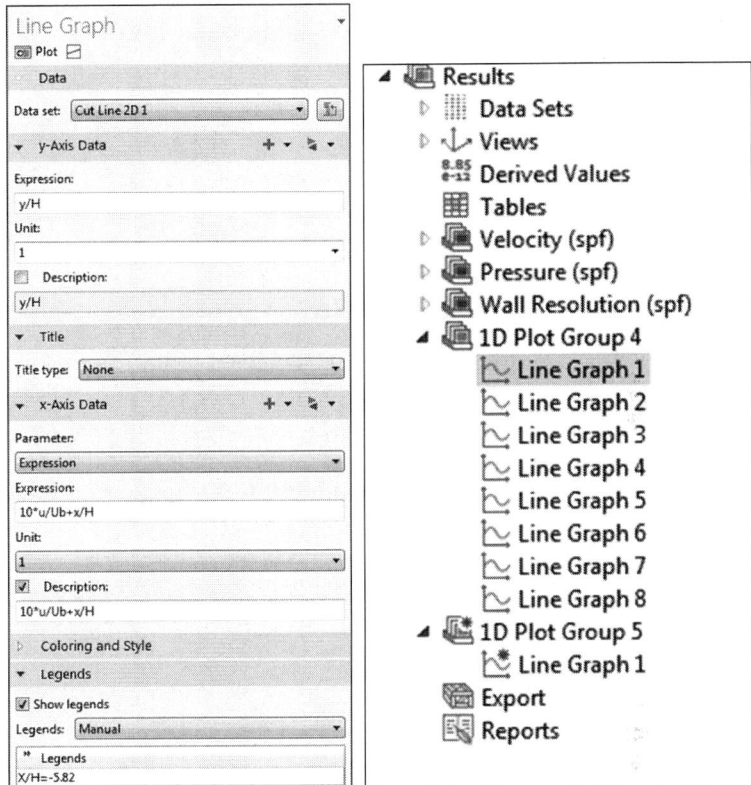

FIGURE 4.12: Window for line graphs axial velocity data entry, data for Line Graph 1 is shown.

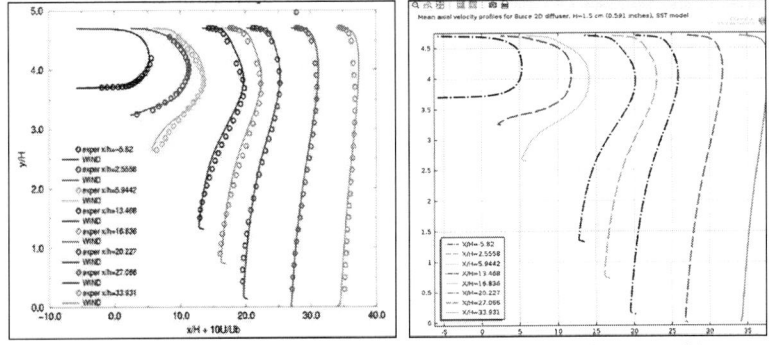

FIGURE 4.13: Model results (right), NASA results (left [Slater 1993]) for mean axial velocity profiles.

13. One of the flow features for the diffuser is the flow detachment and reattachment which result in a recirculation zone. To demonstrate this, we draw streamlines. Right-click on the Velocity node, and select Streamline in the Model Builder window. In the Streamline window, locate the Streamline Positioning section and select Start point controlled from the list for Positioning, Enter 20 for Points, and select Cut Line 2D 5 for Along line. Right-click on the Streamline 1 node and select Color Expression. Click Plot. To see the streamlines more clearly, disable Surface 1 by right-clicking on it and select Disable. The result, as shown in Figure 4.14, clearly shows the recirculation zone, specifically the detachment and re-attachment locations of the separated boundary layer on the ramp of the diffuse.

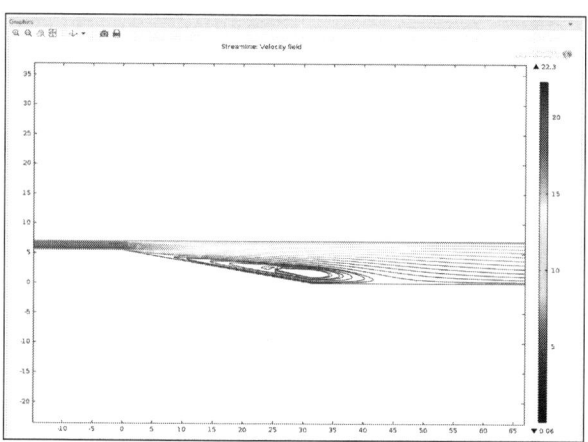

FIGURE 4.14: Streamlines along the diffuser showing the flow separation and re-attachment.

Other quantities like Coefficient of friction, $C_f = \dfrac{\tau_w}{0.5\rho U_b^2}$ and coefficient of pressure $C_p = \dfrac{P - P_{(-1.7)}}{0.5\rho U_b^2}$, for upper and lower walls can be extracted from the model results. $P_{(-1.7)}$ is the static pressure at the point $\dfrac{x}{H} = -1.7$. Results for coefficient of friction are shown in Figure 4.15. Please refer to the built model available on the accompanying disk.

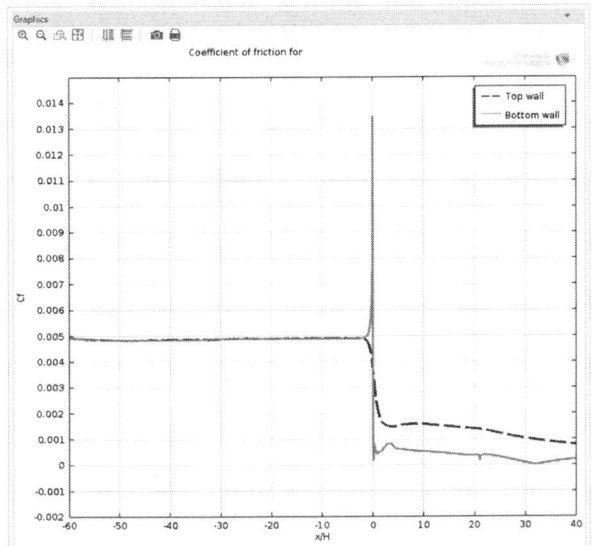

FIGURE 4.15: Coefficient of friction for the diffuser top and bottom walls.

Example 4.2: Modeling of turbulent flow around the S809 airfoil

For this example, we model flow around an airfoil. Airfoils are used in many industries, including aerospace, wind turbine, and turbo-machinery. Proper design and prediction of airfoil characteristics are of major importance for their desired performances. For this example we choose an S-type airfoil, specifically S809, whose aerodynamic performance are representative of typical horizontal-axis wind turbine blades with a rotor diameter of 20–30 m [87]. S809 is a 21%-thick airfoil designed by Somers [88] to achieve maximum lift, low profile drag, and insensitivity to surface roughness. Experimental data for the airfoil section geometry and aerodynamic coefficients are taken from Somers [88] and NREL [89], for the purpose of comparison and validation of the modeling results. Figure 4.16 shows a sketch of the airfoil section. For our example the airfoil chord C = 600 mm.

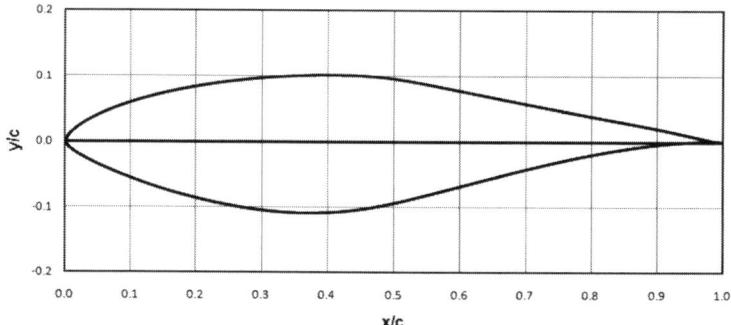

FIGURE 4.16: S809 airfoil profile

We select and use the S-A turbulent model for this example. The flow domain is considered to be about 100 times bigger than the chord in 2D space, in order to minimize the effect of applied boundary conditions on the airfoil performance. All calculations are performed for a chord-based Reynolds number of 2×10^6, which is equivalent to a bulk air speed of 50 m/s with density of 1.204 kg/m^3 and viscosity of 1.806×10^{-5} Pa.s.

Solution:

1. Launch COMSOL and from File>Save as, in the New window, save the model as Example 4.2. Click on the Model Wizard icon.

2. From the Select Space Dimension window, click on the 2D icon. The Select Physics window will appear. Select Fluid Flow>Single-Phase Flow>Turbulent Flow>Turbulent Flow, Spalart-Allmaras (spf). Then click on the Add button.

3. Click on the Study icon/arrow and, from Select Study window, under Preset Studies, click on Stationary with Initialization. Click on the Done icon. The COMSOL Desktop interface appears. In the Geometry window locate the Units section and change the Length unit to mm.

4. Now we make a list of input data used for this model as parameters. From the Home tab ribbon, click on Parameters. In the Parameters window, enter the data (case sensitive) as shown in Figure 4.17. Alternatively this data can be imported from the accompanying disk.

Parameters ▼

▼ Parameters

» Name	Expression	Value	Description
C	600[mm]	0.60000 m	airfoil chord
Re	2e6	2.0000E6	Reynolds number
Air_dens	1.204[kg/m^3]	1.2040 kg/m³	air density
Air_vis	1.806e-5[Pa*s]	1.8060E-5 Pa·s	air viscosity
Ub	Re*Air_vis/(Air_dens*C)	50.000 m/s	bulk inlet velocity
alpha	0	0	angle of attack

FIGURE 4.17: Parameters data for Example 4.2.

5. We build a half-circle for the flow domain covering the up-stream and including the airfoil. Right-click on the Geometry 1 node in the Model Builder window, and select Circle from the list. In the Circle settings window, as shown in Figure 4.18, enter 50*C for Radius and 180 for Sector angle. Choose Center from the Base list and enter C for x. for Rotation, enter 90.

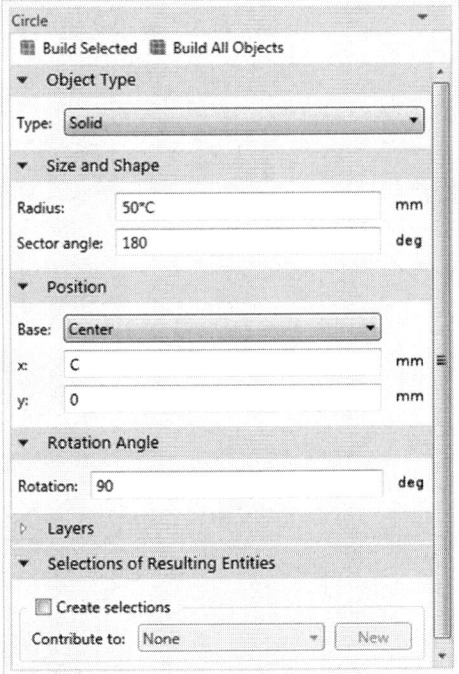

FIGURE 4.18: Data entry for quadrant domain.

6. To build the geometry of the airfoil right-click on the Geometry node, in the model tree, and select Interpolation Curve from the list. In the corresponding window enter the data for the Upper surface as given in Table 4.2 [88]. Alternatively this data can be imported from the accompanying disk. Click on the Build Selected icon. Zoom in to see the curve airfoil upper-surface curve created. Similarly, create another Interpolation Curve and enter data for Lower surface, and then click on Build All Objects. Use the Zoom In button to see the airfoil curves in the Graphics window. Alternatively, users can import the data provided through the accompanying disk into the corresponding tables. The cross section of the airfoil can be built by using these curves. This is done by converting the enclosed area by these curves to a solid. Click on the Geometry tab and select Convert to Solid, from the list under Conversions. In the corresponding window add the upper and lower curves (i.e. ic1 and ic2) into the Input objects by clicking on them in the Graphics window (using the Select Box from the Graphics toolbar might help). Then click on Build Selected.

TABLE 4.2: Coordinates for the Upper and the Lower surfaces of S809 airfoil [88].

Upper surface/curve		Lower surface/curve	
x(mm)	y(mm)	x(mm)	y(mm)
0	0	0	0
0.00037*C	0.00275*C	0.0014*C	−0.00498*C
0.00575*C	0.01166*C	0.00933*C	−0.01272*C
0.01626*C	0.02133*C	0.02321*C	−0.02162*C
0.03158*C	0.03136*C	0.04223*C	−0.03144*C
0.05147*C	0.04143*C	0.06579*C	−0.04199*C
0.07568*C	0.05132*C	0.09325*C	−0.05301*C
0.10390*C	0.06082*C	0.12397*C	−0.06408*C
0.13580*C	0.06972*C	0.15752*C	−0.07467*C
0.17103*C	0.07786*C	0.19362*C	−0.08447*C
0.2092*C	0.08505*C	0.23175*C	−0.09326*C
0.24987*C	0.09113*C	0.27129*C	−0.1006*C
0.29259*C	0.09594*C	0.31188*C	−0.10589*C

Upper surface/curve		Lower surface/curve	
x(mm)	y(mm)	x(mm)	y(mm)
0.33689*C	0.09933*C	0.35328*C	−0.10866*C
0.38223*C	0.10109*C	0.39541*C	−0.10842*C
0.42809*C	0.10101*C	0.43832*C	−0.10484*C
0.47384*C	0.09843*C	0.48234*C	−0.09756*C
0.52005*C	0.09237*C	0.52837*C	−0.08697*C
0.56801*C	0.08356*C	0.57663*C	−0.07442*C
0.61747*C	0.07379*C	0.62649*C	−0.06112*C
0.66718*C	0.06403*C	0.67710*C	−0.04792*C
0.71606*C	0.05462*C	0.72752*C	−0.03558*C
0.76314*C	0.04578*C	0.77668*C	−0.02466*C
0.80756*C	0.03761*C	0.82348*C	−0.01559*C
0.84854*C	0.03017*C	0.86677*C	−0.00859*C
0.88537*C	0.02335*C	0.90545*C	−0.0037*C
0.91763*C	0.01694*C	0.93852*C	−0.00075*C
0.94523*C	0.01101*C	0.96509*C	0.00054*C
0.96799*C	0.006*C	0.98446*C	0.00065*C
0.98528*C	0.00245*C	0.99612*C	0.00024*C
0.99623*C	0.00054*C	1*C	0
1.*C	0	_	_

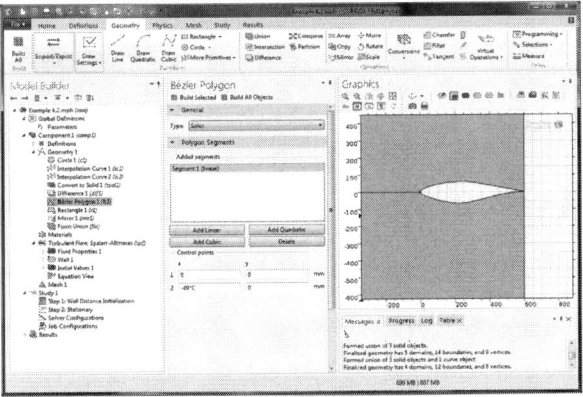

FIGURE 4.19: S809-built geometry with partial flow domain.

7. Now we subtract the airfoil from the semi-circle domain to create the flow domain around the airfoil. Click on Difference, located in the ribbon under the Geometry tab. In the corresponding window select and add domain (c1) to list for Objects to add, and airfoil (*csol1*) to the Objects to subtract (again using the Select Box from the Graphics toolbar might help). Users may have to toggle the Active button to turn it on. Click on Build Selected. For further meshing use, we draw a line to divide the semi-circular flow domain into two sections. Right-click on the Geometry node, in the Model Builder window, and select the Bezier Polygon. In the corresponding settings window, click on the Add Linear button and enter coordinates (0, 0) for control point 1 and (−49*C, 0) for point 2. Click on Build Selected. Figure 4.19 shows the result.

8. To build the remaining of the flow domain geometry for the downstream, right-click the Geometry 1 node in the model tree and select Rectangle from the list. In the corresponding window locate the Size section and enter 100*C for the Width, and 50*C for Height. Locate the Position section and enter *C* for *x*: and click on the Build Selected icon. Click on Mirror in the Geometry tab and in the corresponding window, add Rectangle (r1) to the list and select the Keep input objects check-box. Locate Normal vector to Line of Reflection section and enter 0 for *x*, and 1 for *y*. Click on Build Selected. Click on the Zoom Extents button, located in the Graphics window toolbar, to view the whole flow domain geometry. The final flow domain geometry is shown in Figure 4.20, Zoom-in to see the airfoil in the middle.

9. To set up the inner boundaries for mesh control purposes (a feature to be used for user defined mesh), click on the Geometry tab and select Virtual Operations>Mesh Control Edges. Add all inner boundaries (1, 2, 4, 5) to the Input list. Click the Build Selected button. These boundaries will disappear from the geometry, but will be visible during meshing operation.

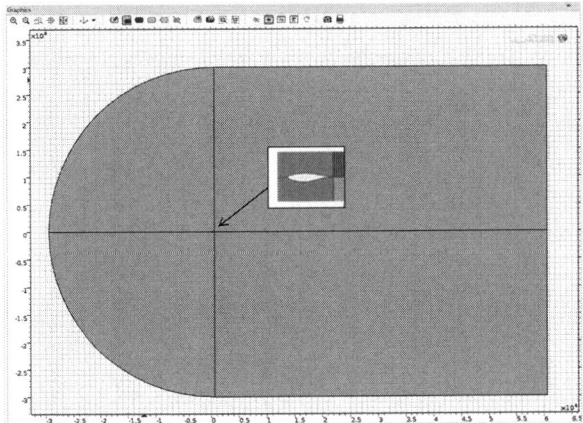

FIGURE 4.20: Built geometry for flow domain around S809 airfoil.

10. Now we specify the fluid properties. Click on the Fluid Properties 1 node, located under Turbulent Flow, Spalart-Allmara (spf). In the corresponding window, locate the Fluid Properties section and select User defined from the list for both Density and Dynamic viscosity and enter Air_dens and Air_vis, respectively. Locate the Distance Equation section to define l_{ref}; this variable is used to control the distance to the wall equation. Any object smaller than its defined value is diminished and only larger objects or walls are considered [55]. For having the airfoil included we assign 0.1 m (< C = 0.6 m) for this variable. We leave the remaining selections from this window as defaults.

11. To set up the physics, click on the Turbulent Flow, Spalart-Allmara (spf) node in the Model Builder window. In the corresponding window, make sure that domain 1 is selected and listed in the Domain Selection section. Users are encouraged to open the Equation section and familiarize themselves with the governing equations (see Chapter 2).

12. To set up flow boundary conditions, click on the Physics tab and select Inlet from the list under Boundaries, located in the ribbon bar. In the Inlet settings window, click on edge (boundary 1) of the flow domain and add them to the list under Boundary Selection. Also locate the Velocity section and select Velocity field option. Enter Ub*cos(alpha*pi/180) and

Ub*sin(alpha*pi/180) for velocity vector **u**, components *x* and *y*, respectively. Note that for these entries we should exactly use the variables as defined in the Parameters (i.e. case sensitive). Similarly, create an open boundary for the outlet, by selecting Open Boundary from the list under Boundaries and assign the right edge (boundary 2) of the flow domain to it. COMSOL, by default, assigns all other boundaries (i.e. airfoil surfaces) as no-slip walls, which can be checked out by clicking on the Wall 1 node, in the Model Builder window.

13. To help the convergence of numerical results we set an initial value for the velocity field. Click on the Initial Values 1 node, and in the corresponding Initial Values settings window locate the Initial Values section. Enter Ub*cos(alpha*pi/180) and Ub*sin(alpha*pi/180) for velocity vector **u**, components *x* and *y*, respectively.

14. To build a mesh, click on the Mesh 1 node in the Model Builder window. In the Mesh window leave the Physics-controlled mesh as a default selection and select Extremely fine option for Element size. Right-click on the Mesh 1 node and select Edit Physics-Induce Sequence, from the list. Click on Boundary Layers 1>Boundary Layer Properties 1. In the corresponding settings window select and add airfoil boundaries (edges 3, 4) to the list. Change the Number of boundary layers to 20 and enter 1.1 for Boundary layer stretching factor. Click on the Boundary Layer 1 node, in the Model Builder window, and locate the Corner Settings section in the corresponding settings window. Select Splitting for Handling of sharp corners. Click on the Build All button. A hybrid mesh consisting of 814 quadrilateral boundary-layer elements and 114,368 triangular elements is built. Figure 4.21 shows the built elements around the leading edge of the airfoil. Note the boundary-layer type elements close to the airfoil surface. A more detailed and comprehensive mesh could be built for further detailed studies.

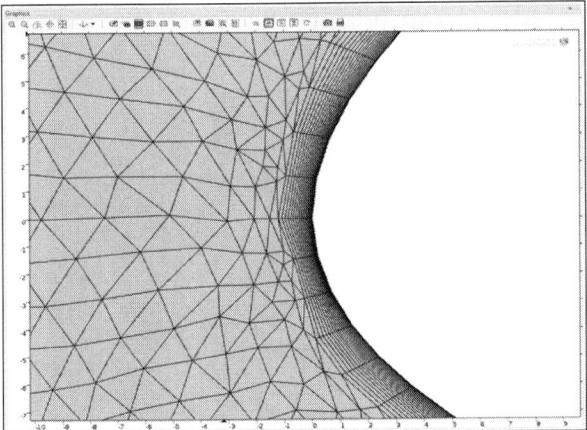

FIGURE 4.21: Hybrid mesh built around the leading edge of the S809 airfoil.

15. To set the model to run for several values of angle of attack, click on the Study 1>Step 2: Stationary node in the Model Builder window. In the Stationary settings window locate and expand Study Extensions (see Figure 4.22). Select the Auxiliary sweep check box. Click add (i.e. plus button located at the bottom of this section) and select alpha (angle of attack), from the list under Auxiliary parameter. Enter range (0,2,14) under the Parameter value list. This will set the model to run for a range of values from 0 to 14 with an increment of 2°, for the angle of attack.

16. It is useful and instructive to plot the solution during the computation. Click on the Study tab from the toolbar and click on Get Initial Value in the ribbon. Expand the Results node in the model Builder window and click on Velocity (spf). From the corresponding settings window locate the Plot Settings section and select View 1 from the list for View. In the Model Builder window click on the View 1 node under Component 1>Definitions, then in the View settings window select the Lock axis check-box. To draw the streamlines, right-click on Velocity (spf) node in the Model Builder window and select Streamline from the list. In the corresponding window locate Streamline Positioning and select Start point controlled from the list for Positioning option. Select Coordinates for Entry method and enter C and range(−600,15,600) for x and y, respectively. With this setting we can plot streamlines close to the airfoil during the solution and for final results (see Figure 4.22).

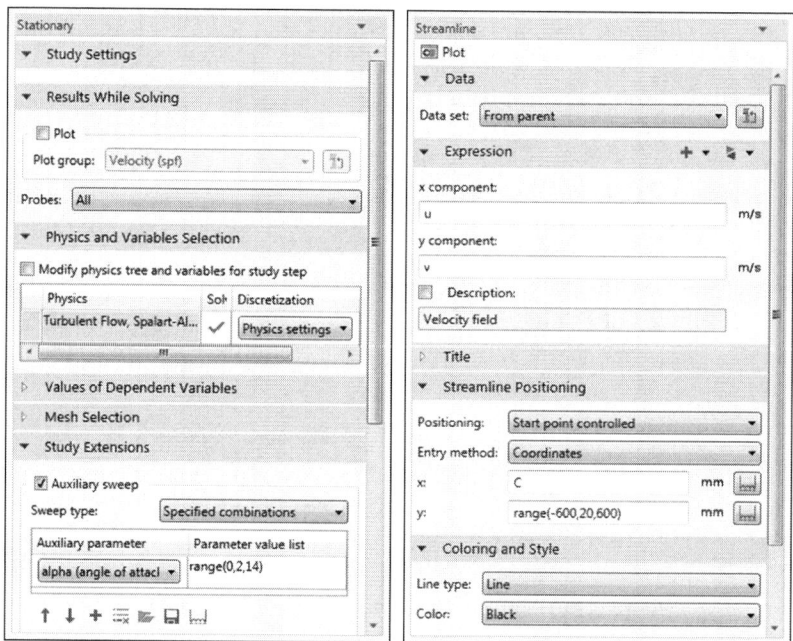

FIGURE 4.22: Data entry for angle of attack solver sweeping parameter and streamline close to the airfoil.

17. To visualize the results during computation, expand Study 1 > Solver Configuration > Solver 1 > Stationary Solver 2 and then click on the Segregated 1 node. In the corresponding Segregated settings window expand Results While Solving and select the Plot check-box.

18. The model is ready for computation. Users may detach the Graphics window by making it Float and by resizing it (using Zoom In/Out tools) to clearly see the streamlines around the airfoil during computation. Click on the Study tab and Compute button ribbon. After computations end, default results are shown for velocity and pressure contours. Figure 4.23 shows the results close to the leading edge for the angle of attack of 8° and Figure 4.24 shows the results for the whole airfoil at 14°.

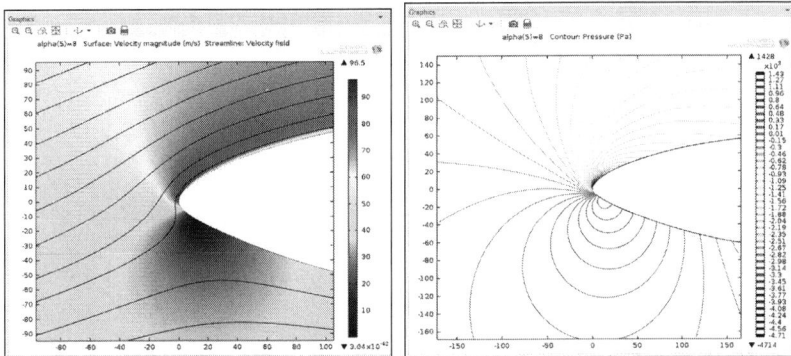

FIGURE 4.23: Velocity, streamlines, and pressure contour close to the leading edge at angle of attack of $8°$.

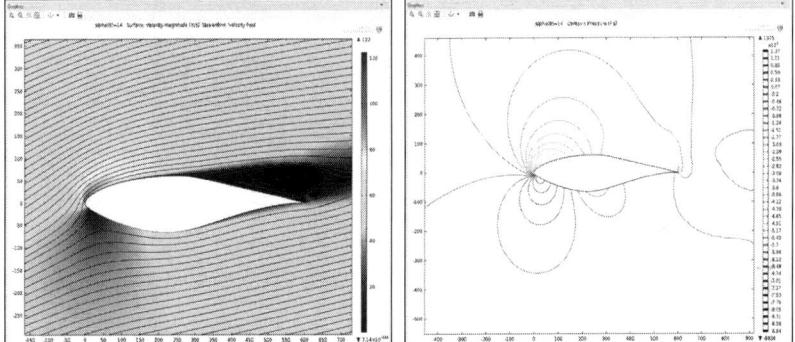

FIGURE 4.24: Velocity, streamlines, and pressure contour around S809 airfoil at angle of at attack of $14°$.

To verify the results, we calculate the drag, lift, and pressure coefficients. These coefficients are defined as

$$C_p = \frac{-\oint p}{C(0.5 \times \text{Air\_dens} \times U_b^2)}$$

$$C_D = \frac{-\oint p}{C(0.5 \times \text{Air\_dens} \times U_b^2)} \times [ny.\sin(\alpha) + nx.\cos(\alpha)]$$

$$C_L = \frac{-\oint p}{C(0.5 \times \text{Air\_dens} \times U_b^2)} \times [ny.\cos(\alpha) + nx.\sin(\alpha)]$$

where *nx* and *ny* are components of element surface normal vector, as defined in the COMSOL manual by nxmesh and nymesh, respectively. Line integrals are taken over the airfoil boundary curves.

19. Right-click on Results > Derived Values and select Integration>Line Integration. In the corresponding settings window, select and add airfoil curves (curves 3, 4) to the Selection list. Locate the Expression section and type in p/(0.5*Air_dens*Ub^2)/C*(spf.nymesh*sin(alpha*pi/180)+spf.nxmesh*cos(alpha*pi/180)). Select the Description check box and type in Drag coeff. Cd. Click on the Evaluate button. Results for drag coefficients for all values of angle of attack will appear in the Table 2 window. In the Table window toolbar, click on the Table Graph icon. The results will appear as a graph in the Graphics window. In the Table Graph window, locate the Line markers section and select Point for Marker and In data points for Positioning. Click the Plot button. Figure 4.25 shows the model results.

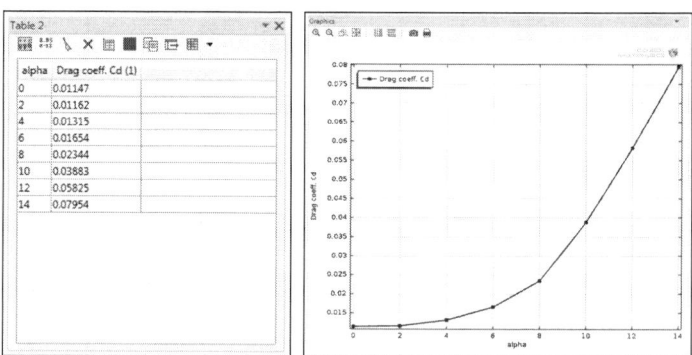

FIGURE 4.25: Model results for Drag coefficients of airfoil S809 versus angle of attack.

20. Repeat Instructions given in step 19 in order to create another line integrations and graph for Lift coefficient. For Lift type in the following expression in the Expression section; p/(0.5*Air_dens*Ub^2)/C*(spf.nymesh*cos(alpha*pi/180)-spf.nxmesh*sin(alpha*pi/180)).
Results are shown in Figure 4.26.

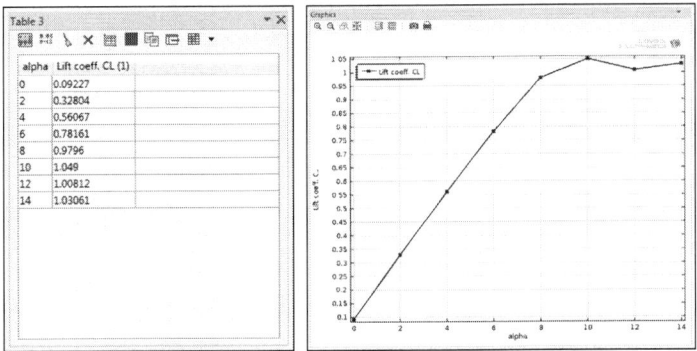

FIGURE 4.26: Model results for Lift coefficients of airfoil S809 versus angle of attack.

21. Repeat Instructions given in step 19, to create another line integration and graph for pressure coefficient. For pressure, type the following expression into the Expression section: $(-p/(0.5^*\text{Air\_dens}^*\text{Ub}^{\wedge}2))/C$. Results are shown in Figure 4.27.

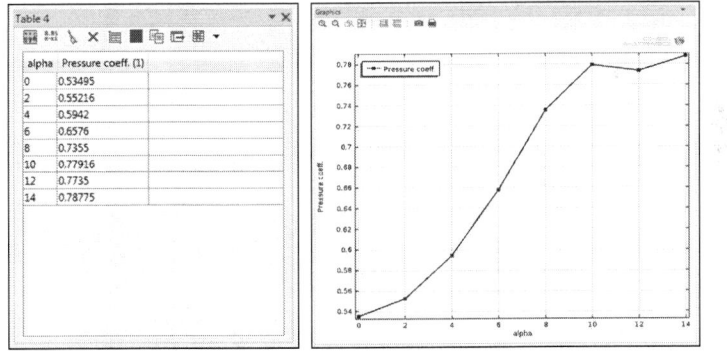

FIGURE 4.27: Model results for pressure coefficients of airfoil S809 versus angle of attack.

22. Some experimental results used by Wolfe and Ochs [90]. Comparable modeling results are given in Table 4.3, using linear interpolation for the values of α given. A finer user-defined custom mesh seems suitable for more accurate results.

TABLE 4.3: Experimental [90] values of drag and lift coefficients. Compared with model results for S809 airfoil.

α, angle of attack (deg)	CL		Cd	
	Model*	experiment [90]	Model*	experiment [90]
0	0.0923	0.1469	0.01147	0.007
1.02	0.095	0.2716	0.01147	0.0072
5.13	0.877	0.7609	0.015	0.007
9.22	1.022	1.0385	0.0328	0.024
14.24	1.033	1.1104	0.082	0.09

23. It is also useful to compare the pressure coefficient at a given angle of attack over the airfoil boundary. Click on the Results tab and select 1D Plot Group. From the ribbon select Line Graph. In the Line Graph settings window, select and add airfoil boundaries (3, 4) to the Selection list. Locate the Expression section and type in p/(0.5*Air_dens*Ub^2). Locate the Data section and select Solution 1 from the list. Select 'From list' from the selections for Parameter selection (alpha) and click on 0 in the list. Locate x-Axis Data section and choose Expression from the list under Parameter. Type in x/c for the Expression. Locate the Legend section and select the Show legend check box. Duplicate Line Graph 1. The node Line Graph 2 will be created. In the corresponding window, modify the settings for angle of attack equal to 14, by selection 14 from the list under Parameter values (alpha). Click on Plot. Results are shown in Figure 4.28.

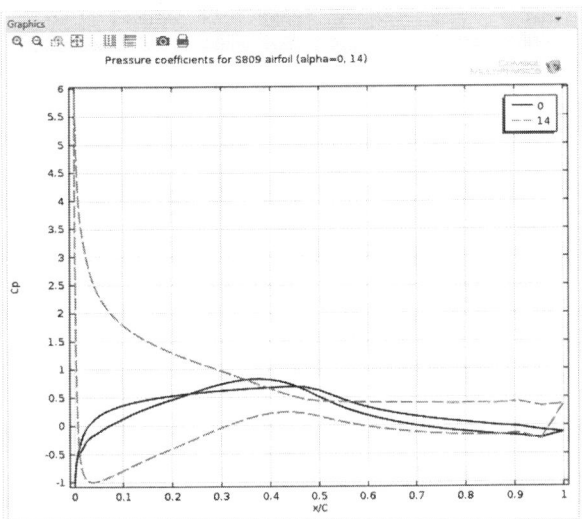

FIGURE 4.28: Pressure coefficients at two values of angle of attacks (0, 14) for airfoil S809.

Example 4.3: Modeling of turbulent flow in a pipe with 90° bend

In this example we model the turbulent flow through a pipe with a 90° bend, as shown schematically in Figure 4.29. The introduction of a 'simple' bend in a straight pipe makes the flow pattern more complex and has physical consequences like excessive corrosion and erosion. Examples of this type of flow can be found in human arteries, rivers, industrial pipe networks, and internal combustion engines. The Reynolds number Re is a parameter which can be interpreted as the ratio of momentum flux (i.e. inertial forces per unit area) over the wall shear stress (i.e. viscous forces), perhaps in the straight section of a pipe with diameter D. When the pipe is curved (with radius of curvature R_c in the plane of the pipe) then the centripetal acceleration changes the balance of acting forces (i.e. inertia, viscous and centrifugal) on the fluid and hence different flow pattern develops through the bend, which has upstream and downstream effects, as compared to the straight section of the pipe. Among the effects is the development of secondary flows at the bend which are demonstrated by two counter-rotating vortices at a plane perpendicular to the center-line axis of the pipe. Dean [91] discovered the secondary flow, for a laminar flow, and also introduced a parameter that dynamically defines such flows, i.e. Dean number $D_e = R_e\sqrt{\dfrac{D}{2R_c}}.$

Dean number could be interpreted as the ratio of centripetal and inertia forces over viscous forces.

For the model input data and validation we use the data reported by Homicz [92], who used the k-ε turbulent model for his model of this problem. In this example we use the k-ω model, to capture the flow separation and curvature, relatively more accurately. COMSOL has provided a similar model,[13] available to registered users through their website. However, we don't exactly follow the solution provided by COMSOL for this model example.

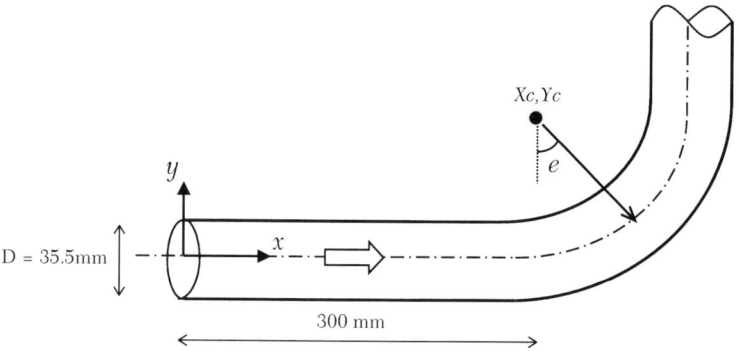

FIGURE 4.29: The pipe bend geometry and dimensions (sketch).

Solution:

1. Launch COMSOL and from File>Save as, in the New window, save the model as Example 4.3. Click on the Model Wizard icon.

2. From the Select Space Dimension window, click on the 3D icon. The Select Physics window will appear. Select Fluid Flow>Single-Phase Flow>Turbulent Flow>Turbulent Flow, k-ω *(spf). Then click on the Add button.*

3. Click on the Study icon/arrow and, from the Select Study window, under Preset Studies, click on Stationary. Click on the Done icon. The COMSOL Desktop interface will appear. In the Geometry window locate the Units section and change the Length unit to mm.

---

[13] CFD_Module/Single-Phase_Benchmarks/pipe elbow.

4. Now we make a list of input data used for this model as parameters. From the Home tab ribbon, click on Parameters. In the Parameters window, enter the data (case sensitive) as shown in Figure 4.30. Alternatively this data can be imported from the accompanying disk.

Name	Expression	Value	Description
D	35.5[mm]	0.035500 m	pipe diameter
Rc	50[mm]	0.050000 m	bend radius of curvature
Fluid_dens	965.35[kg/m^3]	965.35 kg/m³	fluid density
Fluid_vis	3.14e-4[Pa*s]	3.1400E-4 Pa·s	fluid viscosity
Ub	5[m/s]	5.0000 m/s	bulk speed
L	200[mm]	0.20000 m	length scale
Re	Fluid_dens*Ub*D/Fluid_vis	5.4570E5	Reynolds no.
I_turb	0.16/Re^(1/8)	0.030690	turbulece intensity
L_turb	0.07*D/2	0.0012425 m	turbulence length scale

FIGURE 4.30: Parameters data for Example 4.3.

We use drawing tools available in COMSOL for building the pipe geometry. However, users may want to build it in their own favorite CAD software and import it into this model. For reference we place the origin of the coordinates system at the center of inlet of the pipe, as shown in Figure 4.29. Due to geometry and flow symmetries we only model the upper half of the pipe. Hence the symmetry plane lies on the plane at $z = 0$.

5. We build a half-circle for the inlet of the pipe. Right-click on the Geometry 1 node, in the Model Builder window, and select Work Plane. In the Work Plane settings window, locate the Plane Definition section and select yz-plane from the list for Plane. Right-click on the Plane Geometry node, in the Model Builder window, and select Circle from the list. In the Circle settings window, enter $D/2$ for Radius and 180 for Sector angle. Click on the Build Selected icon. To build the section of the pipe upstream of the bend, we extrude the half-circle, just built. From the ribbon under Geometry tab, click on Extrude. In the Extrude settings window locate Distances from Plane section and enter 1.5*L under Distances (m). Click on the Build Selected icon.

6. To build the pipe bend, we use the Revolve tool. Click on Work Plane in the ribbon toolbar, under the Geometry tab.

In the Work Plane settings window, locate the Plane Definition section and select Coordinates from the list for Plane type. Enter the following coordinates for Point 1: (1.5*L,0,0), for Point 2: (1.5*L,D/2,0), and for Point 3: (1.5*L,0,D/2). Right-click on the Plane Geometry node, under Work Plane 2, and Select Circle from the list. In the Circle settings window, enter D/2 for Radius and 180 for Sector angle. Right-click on the Work Plane 2 node in the Model Builder window, and select Revolve. In the Revolve settings window, locate the Revolution Angles section and enter 0 and 90 for Start angle and End angle, respectively. In the Revolution Axis section, select 3D for Axis type and enter following coordinates data. Click on Build All Objects.

▼ Revolution Axis		
Axis type:	3D	▼
Point on the revolution axis		
x:	1.5*L	m
y:	Rc	m
z:	0	m
Direction of revolution axis		
x:	0	
y:	0	
z:	1	

7. Similarly, we build the last section of the pipe, downstream of the bend. Right-click on the Geometry 1 node in the Model Builder window, and select Work Plane. In the Work Plane window, locate the Plane Definition section and select Face Parallel for Plane type. Add the semi-circle face of the bend (rev1-6) to the Plane face list. Locate and expand the Local Coordinate System section and select Vertex projection for Origin, then add the origin of half-circle from the bend section to the Vertex for origin. Right-click on Plane Geometry, under the Work Plane 3 node, and select Circle. Enter D/2 for its Radius and 180 for Sector angle. Click the Build Selected icon. Right-click on the Work Plane 3 node and select Extrude. In the Extrude window, locate the Distance from Plane section and enter L under Distances (mm). Click on the Build All Objects icon.

Final geometry of the flow domain, using symmetry, is show in Figure 4.31.

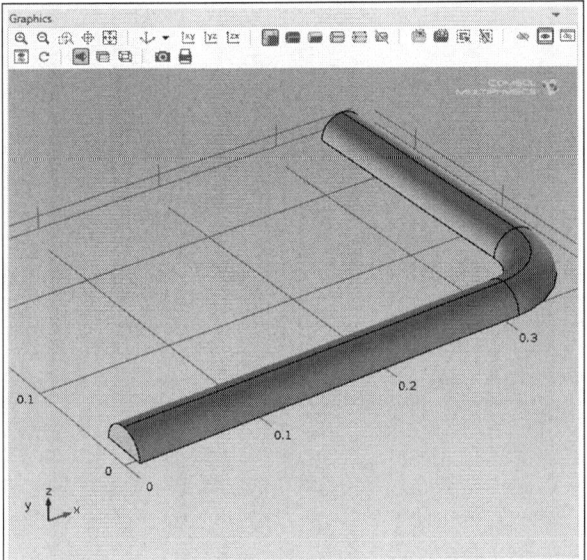

FIGURE 4.31: Upper half of the pipe geometry as built in COMSOL.

8. To set up the fluid properties, click on the Fluid Properties 1 node, in the Model Builder window, and in the Fluid Properties settings window locate the Fluid Properties section. Select User defined for both Density and Dynamic viscosity and enter Fluid_dens and Fluid_vis, for ρ and μ, respectively.

9. To apply boundary conditions, click on the Boundaries button located in the ribbon under the Physics tab. From the list, select Inlet. In the Inlet settings window, select boundary 1 (i.e. pipe entrance) and add it to the list under the Selection section. Locate the Velocity section and enter Ub for U_0. We should also specify the turbulence intensity and length scale. In the Inlet settings window, locate the Boundary Condition section and enter I_turb and L_turb (as the same format as defined in the Parameters) for I_T and L_T, respectively. Similarly, create an Outlet boundary and assign boundary 14 (i.e. pipe exit) to it. In the Outlet settings window, locate the Pressure Conditions section and enter

20[bar] for P_0. To apply the symmetry, click on the Boundary button again and select Symmetry. In the Symmetry settings window, add all boundaries (2, 4, 7, 9, 11, 15) of the pipe located at flat bottom surface of the pipe (i.e. at $z = 0$) to the Selection list. We also define initial values for easing the convergence of the numerical solutions. Click on the Initial Values 1 node in the Model Builder window. In the Initial Values settings window, locate the Initial Values section and enter 20.1[bar] for P.

The physics of the model is complete at this stage of the model building process. Now we should create a mesh. For this purpose, we start with automatic meshing tool available in COMSOL and determine a user-defined mesh as an exercise problem that users can built then compare the results. Homicz's report [92] suggests a mesh which contains more than half a million elements. Another issue for this mesh, since this model is 3D, is the high demand for computer power, both RAM and CPU. In COMSOL there is a tool (i.e. Update Solution) to ease on this requirement, with which we can solve the model using a relatively coarser mesh and then update the solution for a finer mesh using interpolation technique. This way we introduce some relative approximation to the solution, as compared to solving the model for a finer mesh, but it saves us computer power. If users have access to a 'workstation'-type computer, we recommend solving this model with Extra fine or Extremely fine mesh, otherwise use the method presented in the following steps.

10. Click on the Mesh 1 node, in the Model Builder window, and from the Mesh settings window select Fine for Element size. Click on the Build All button. This creates a total of more than 600 thousand volume elements. A close-up of the mesh at the pipe entrance and near the bend are shown in Figure 4.32.

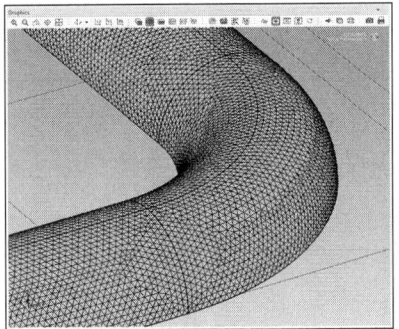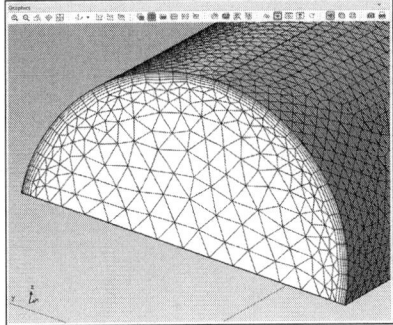

FIGURE 4.32: A Fine-resolution hybrid mesh built for the pipe, left) surface elements around the bend, right) close-up at the inlet.

11. To run the model click on the Study tab button and select Compute from the ribbon. It takes a while for computations to finish (about 3 hours on a typical computer). The first result quantity to check is the Wall resolution (see Chapter 2). This quantity is the dimensionless distance from the wall and COMSOL suggests that it should be much larger than 11.06. Click on the Wall Resolution (Spf) node, in the Model Builder window. The values of variable δ_W^+ (i.e. spf.d_w_plus) are shown in the Graphics window. The maximum value is about 60. This seems high and we should resolve the mesh and run the model again, using the resolved mesh. However, we create an Extra fine mesh but instead of running the model using this mesh (and merely for saving computer power and time) we use an interpolation technique for finding the solution using the existing solution (i.e. the results for Fine mesh, already obtained). Click on the Mesh 1 node, in the Model Builder window and from the Mesh window select Extra fine for Element size. Click on Build All button. A mesh which consists of more than 3 million elements is built. Now click on the Study tab and from the corresponding ribbon list click on the Update Solution button. After results become available, click on the Wall Resolution (Spf) node in the Model Builder window. The values of variable δ_W^+ (i.e. spf.d_w_plus) are shown in Graphics window. The maximum value is about 34. We accept this value (users may want to improve on this or, as mentioned, run the model using the refined mesh) and move forward to build more graphs using the results.

12. The default results show the Velocity and pressure contours. We would like to show the velocity magnitude on horizontal surfaces, parallel to the symmetry plane of the pipe. Click on the Slice 1 under Velocity (spf) node in the Model Builder window. Right-click on it and rename it Slice 1 at z = 0. In the Slice settings window locate the Title section and select None from the list for Title type. Also locate the Plane Data section and select xy-planes for Plane, and Coordinates for Entry method. Enter 0 for z-coordinates. In order to show the results at the bend and downstream from it, we create a filter. Right-click on the Slice 1 at z = 0 node, in the Model Builder window, and select Filter from the list. In the Filter settings window type in $x>1.5^*L$ for the Logical expression for inclusion. For the purpose of visualization we move the slice downward. Right-click on the Slice 1 at z = 0 node and select Deformation from the list. In the Deformation settings window locate the Expression section and enter -1[mm] for x component and -1[mm] for z component. Check the Scale factor box and enter 20. Similarly, build two more slices at z = $D/4$ and z = $3D/8$, respectively. The latter is displaced by 1[mm] for x component and 1[mm] for z component. The final velocity graphs are shown in Figure 4.33.

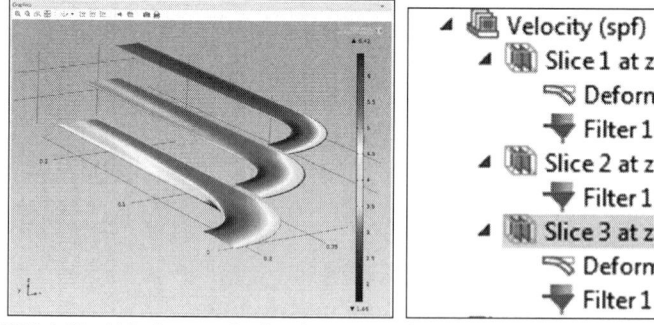

FIGURE 4.33: Velocity magnitudes shown at surface z = 0 (bottom), z = $D/4$ (middle), and z = $3D/8$ (top) at and downstream of the pipe bend.

13. Contours of axial velocity at the entrance of the bend ($\theta = 0$), at its mid-plane ($\theta = 45°$), and at its exit ($\theta = 90°$) are desirable results. For building these results we first extract corresponding values from the solution. Click on the Results tab and then click on Cut Plane, located in the Data Set

group in the ribbon. In the Model Builder window, rename Cut Plane 1 node to Cut Plane 1-bend entrance. In the Cut Plane settings window locate the Plane Data section and select yz-planes for Plane and enter 1.5*L for x-coordinate. Similarly create another cut plane and rename it Cut Plane 2-bend middle. In the corresponding Cut Plane settings, modify the relevant values according to Figure 4.34. The cut plane at the exit can be built similarly, by setting a xz-plane located at R_c for its y-coordinate value, as shown in Figure 4.34.

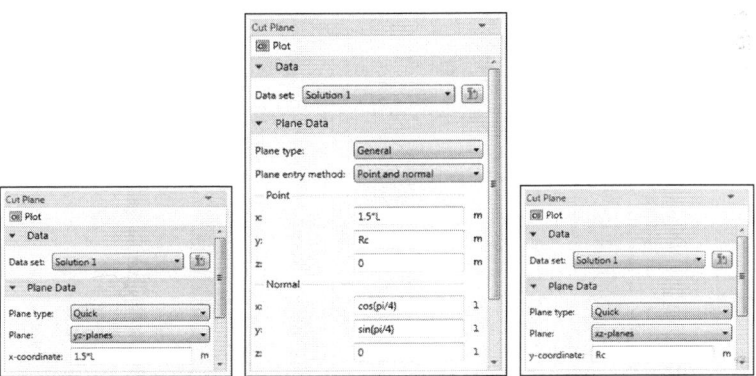

FIGURE 4.34: Settings for building three planes at the entrance (left); mid-section (middle), and exit (right) of the pipe bend.

Now we use these extracted data sets for visualization the axial velocity contours.

14. Click on the Results tab and then on the 3D Plot Group button, located in the Plot Group. Right-click on the 3D Plot Group 4 node, in the Model Builder window, and select Contour from the list. Rename the Contour 1 node Contour 1-bend entrance. In the Contour settings window, locate the Data section and select Cut Plane1-bend entrance, from the list. Type u, for Expression. To scale the legend according to the velocity magnitude, right-click on Contour 1-bend entrance node and select Color Expression. For the purpose of merely clear visualization we displace the contour graph. Right-click on the Contour 1-bend entrance node and select Deformation. In the Deformation settings window, enter -1 [mm] for x component and 20 for the Scale factor, by clicking

on the Scale factor check box. Similarly, create two more Contours for the bend mid-section and bend exit using their corresponding cut planes. Users may edit the legends and titles for each graph. Final contours are shown in Figure 4.35.

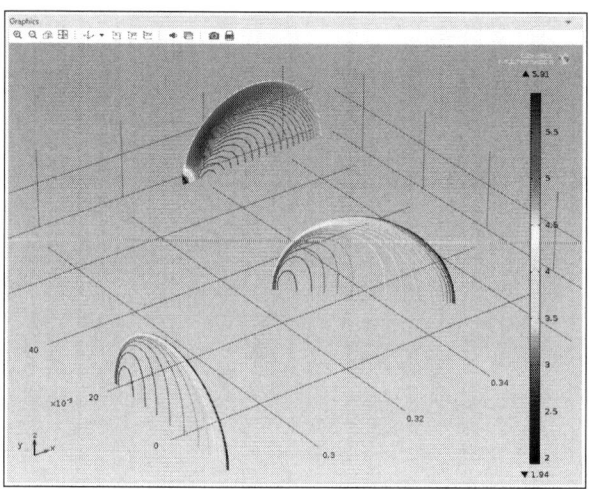

FIGURE 4.35: Contours of axial velocity at the entrance ($\theta = 0°$), middle ($\theta = 45°$), and exit ($\theta = 90°$) of the pipe bend.

A quantity of interest is the so-called diametrical pressure coefficient, c_k [93], defined as

$$c_k = \frac{\bar{p}_o - \bar{p}_i}{0.5\rho U_b^2}$$

where $\bar{p}_o$ and $\bar{p}_i$ are the pressures at the points of intersection of the symmetry plane and the mid-section plane of the pipe bend at the outer and inner radii, respectively. Another relation for c_k could be derived using Dean number D_e;

$$c_k = 4\left(\frac{D_e}{R_e}\right)^2$$

The experimental data are approximately correlated by $c_k \cong 2D / R_c$, for the range of the Reynolds number larger than 5×10^5, [93]. Therefore for our model ($Re = 5.457\ 5 \times 10^5$) we have a value of $c_k \cong \dfrac{2 \times 35.5}{50} = 1.42$. We make use of this value in order to validate our model. We should also mention that Homicz [92] reports a value of c_k = 1.36 and COMSOL a value of c_k = 1.56 for their respective modeling results.

15. We create two points for extracting the values of pressures $\bar{p}_o$ and $\bar{p}_i$ Click on the Geometry tab, expand the More Primitives button, and select Point from the list. Rename the Point 1 node, created under Geometry 1 in the Model Builder window, to Point 1-Pi. In the Point window enter 1.5*L+(Rc-D/2)*cos(pi/4) for x, and Rc-(Rc-D/2)*sin(pi/4) for y. Similarly create another point, rename it Point 2-Po, and enter 1.5*L+(Rc-D/2)*cos(pi/4)+D*cos(pi/4) for x, and Rc-(Rc-D/2)*sin(pi/4)-D*cos(pi/4) for y. For both of these two points z = 0. lick the Build All Objects tab. Click on the Study button and then click on the Update Solution button in the ribbon.

16. Extract the pressure values at the two points created in step 15, click on the Results tab, and then on the Point Evaluation button in the Derived Values group. In the Point Evaluation settings window, select the two points (9 and 12) and add them to the list under the Selection section. Enter p for Expression and click on Evaluate. The values of the pressures appear in the Table 1 window as $\bar{p}_o$ = pressure at point 12 = 2.00781 Mpa and $\bar{p}_i$ = pressure at point 9 = 1.988 Mpa. Using these values we get $c_k = \dfrac{(2.00781 - 1.988) \times 10^6}{0.5 \times 965.35 \times 5^2} \cong 1.642$.

We accept this value as a close approximation of the experimental value of 1.42, in view of the fact that some scattering in experimental data is noted for sharp bends (i.e. $Rc/D > 2$) [93]. However, the accuracy of the modeling results can be improved by refining the mesh and run the model again.

17. It is useful to visualize the streamlines, specifically close to the bend. Click on the Results tab and select the 3D Plot Group from the ribbon in the Plot Group. In the ribbon for 3D Plot Group, click on Streamline. In the Streamline settings window locate the Selection section, then select and add inlet surface of the pipe (boundary 1) to the list. Locate the Coloring and Style section and select Tube for Line type.

In order to set the color scheme based on velocity magnitudes, right-click on the Streamline 1 node and select Color Expressions. The result for streamlines is shown in Figure 4.36.

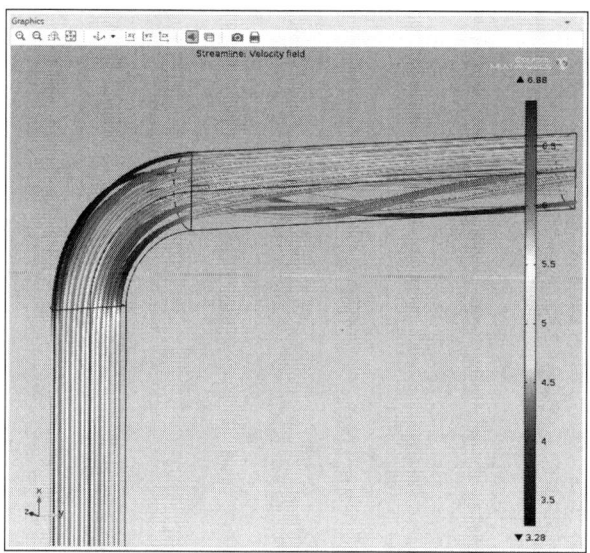

FIGURE 4.36: Streamlines close to the pipe bend, color scaled with velocity magnitude.

Several graphs could be built based on the objectives for a study, for example, pressure contours at the bend are superimposed on the pressure surface values, as shown in Figure 4.37.

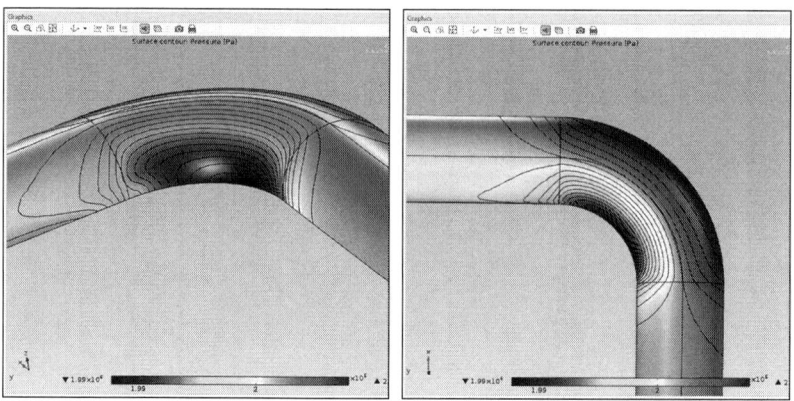

FIGURE 4.37: Pressure contours on the surface of the pipe bend; inside view (left), plan view (right).

Finally we plot the pressure and velocity magnitude on the mid-section plane at the bend for the whole pipe, with using the existing results.

18. Click on the Results tab and from the Data Set group expand the More Data Sets list and select Mirror 3D. In the Mirror 3D window, locate Plane Data and select xy-planes for Plane. This operation will create a data set which is a 'mirror image' of the solution with respect to the symmetry plane. To get the solution for whole pipe at the bend mid-section plane click on the Cut Plane 2-bend middle node, located under the Data Sets node in the Model Builder window, and from the corresponding Cut Plane window select Mirror 3D 1 for the Data set. Now we can use this data for visualization on a surface.

19. Create another 3D Plot Group (similar procedure as explained in above mentioned step 17) and assign Mirror 3D 1 for its Data set. Create a Surface plot and assign Cut Plane 2-bend middle for its Data set, and type in p for its Expression. Also create a Contour plot and again assign Cut Plane 2-bend middle for its Data set, and type in p for its Expression. Click the Plot icon button. Similarly, create a surface plot and a contour plot using velocity magnitude spf.U. Results are shown in Figure 4.38.

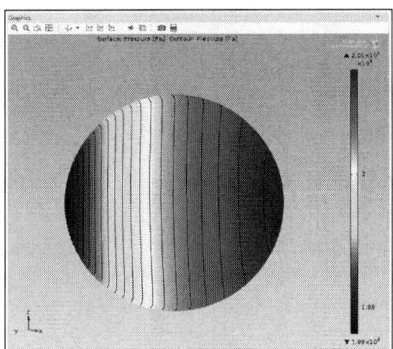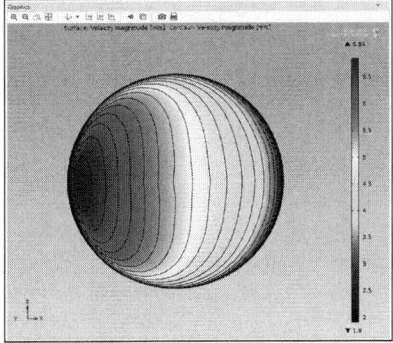

FIGURE 4.38: Pressure (left) and velocity magnitude (right) surface and contour plots at the pipe bend mid-section plane ($\theta = 45°$).

Example 4.4: Modeling of turbulent flow in a ventilated room

Many industrial and residential space heating and ventilation systems are displacement type. For displacement ventilation air enters the room from the floor level and pushes the staled air out of the space. The heat source may be electric or the warmed up fresh air entering the room. The variations of temperature and velocity field are important design parameters for the ventilation system and comfort of the room occupants. In this example we model the turbulent flow and temperature variation in a typical-sized ventilated room.

Flow in a heated room could be forced convection, natural convection, or a combination of both. For an iso-thermal flow the ratio of inertia and viscous forces determines the flow status. That is, for high or low Reynolds number. When air is heated its density changes and buoyancy force is exerted on it. Two dimensionless numbers, Reynolds and Grashof, are usually used for identifying the flow status. The Grashof number $Gr = \dfrac{g\alpha\Delta T L^3}{v^2}$, can be interpreted as the ratio of buoyancy force over viscous force. Similarly, the Reynolds number $Re = \dfrac{UL}{v}$ can be interpreted as the ratio of inertia force over viscous force. Where g is gravitational acceleration, U and L velocity and length scales, respectively, α volumetric thermal expansion coefficient ($\sim 1/T$, for ideal gases), ΔT temperature difference, and v kinematic viscosity of the fluid. We may define a new combined dimensionless number $GRe = \dfrac{Gr}{Re^2} = \dfrac{g\alpha\Delta T}{U^2/L}$, which is the key dimensionless parameter that can be interpreted as the ratio of buoyancy force over inertia force. When GRe is less than unity then the buoyancy forces are negligible and natural convection is weak or negligible and Reynolds number should be used for identifying the laminar or turbulence conditions. But when GRe is larger than unity, buoyancy force dominates and the Grash of number should be used for identifying the laminar or turbulence conditions. We will use these dimensionless numbers to investigate the flow status in the ventilated room.

For this model example we rebuild a relevant model available in the COMSOL Library (i.e. CFD_Module/Non-Isothermal_Flow/

displacement_ventilation)[14]. We modify, to some extent, the COM-SOL model's building sequences including parameters, features, the result analysis, and post-processing.

The model geometry is a room with 3m height, 2.5m depth, and 3.65m width. A warm stream of air (0.028 m³/s, 45°C) enters the room from an inlet located at the floor center. Fresh air (0.05 m³/s, 21°C) is supplied through a wall inlet. Heated air exits the room through an exhaust located in the center of the ceiling. The walls of the room are almost perfectly insulated. Symmetry of the geometry and flow boundary conditions reduces the modeling domain to half of the original size, as shown/depicted in Figure 4.39.

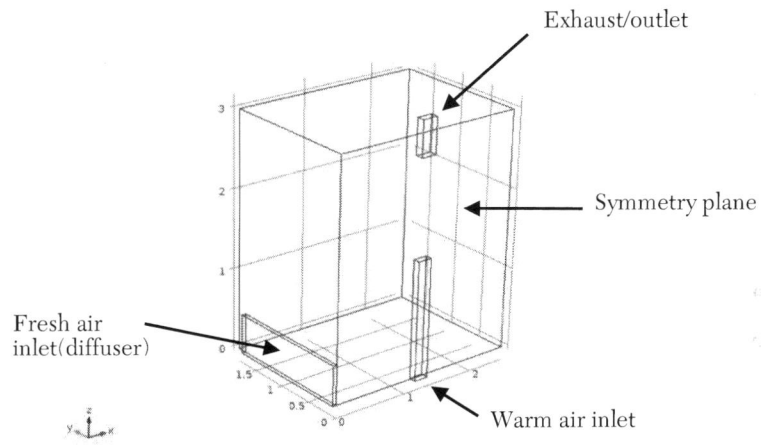

FIGURE 4.39: Room dimensions and modeling domain geometry using symmetry.

For the given data, $GRe = \dfrac{9.81 \times (\frac{1}{300}) \times 20}{1^2 / 2} = 1.308 > 1,$ therefore buoyancy force dominates and the turbulence status is define by the Grashof number, $Gr = \dfrac{9.81 \times (\frac{1}{300}) \times 20 \times 2^3}{(1.6 \times 10^{-5})^2} \cong 2.044 \times 10^{10}.$ This surely indicates a turbulent air flow in the room.

[14.] Model made using COMSOL Multiphysics®, and is provided courtesy of COM-SOL. COMSOL materials are provided "as is" without any representations or warranties of any kind including, but not limited to, any implied warranties of merchantability, fitness for a particular purpose, or noninfringement.

Solution:

1. Launch COMSOL and from File>Save as, in the New window, save the model as Example 4.4. Click on the Model Wizard icon.

2. From Select Space Dimension window, click on the 3D icon. The Select Physics window will appear. Select Fluid Flow > Non-Isothermal Flow > Turbulent Flow > Turbulent Flow, k-ω (nitf). Then click on the Add icon.

3. Click on the Study icon/arrow and, from the Select Study window, under Preset Studies click on Stationary. Click on the Done icon. The COMSOL Desktop interface appears.

4. Now we make a list of input data used for this model as parameters, mainly geometry dependent. From the Home tab ribbon, click on Parameters. In the Parameters window, enter the data (case sensitive) as shown in Figure 4.40. Alternatively this data can be imported from the accompanying disk.

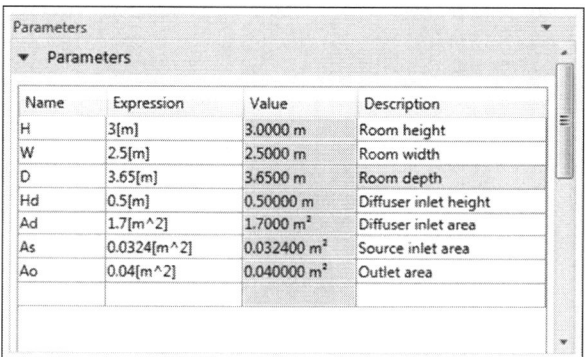

FIGURE 4.40: Parameters data for Example 4.4.

5. We define some variables, which are independent of the room geometry. Right-click on the Component 1 (comp1) > Definitions node, in the model tree, and select Variables. In the Variables settings window, enter the data (case sensitive) as shown in Figure 4.41. Alternatively this data can be imported from the accompanying disk.

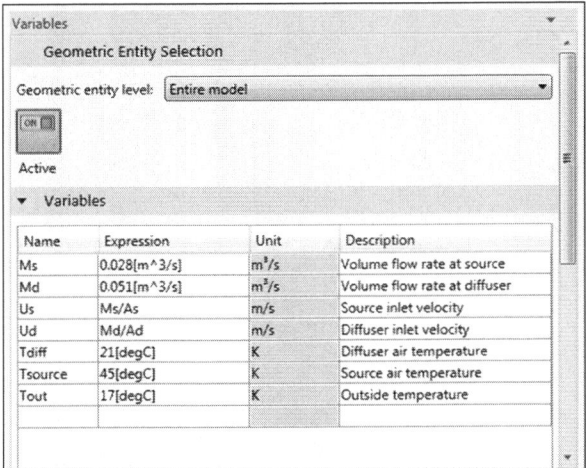

Name	Expression	Unit	Description
Ms	0.028[m^3/s]	m³/s	Volume flow rate at source
Md	0.051[m^3/s]	m³/s	Volume flow rate at diffuser
Us	Ms/As	m/s	Source inlet velocity
Ud	Md/Ad	m/s	Diffuser inlet velocity
Tdiff	21[degC]	K	Diffuser air temperature
Tsource	45[degC]	K	Source air temperature
Tout	17[degC]	K	Outside temperature

FIGURE 4.41: Variables list for Example 4.4.

Now we build the geometry of the computational domain, including the ventilation inlets and outlet. For meshing purposes and minimize the computation time we make use of symmetry and divide the room size in half, as shown in Figure 4.39.

6. From the Geometry tab toolbar select Block. In the Block settings window locate the Size section and enter the data as shown in Figure 4.42 (left picture). This block is the main room space. Similarly, build another Block and enter the dimensions as shown in Figure 4.42 (right picture). Rename these Blocks by right-clicking on the Block1 and Block2, in the Model Builder window, to room and inlet-fresh air, respectively. To combine these two domains, click on Union in the Geometry tab and select both domains (*blk1* and *blk2*) and add them to the list in the Union window. Clear the Keep interior boundaries check box (an important step for building a better mesh later on) and click on Build Selected.

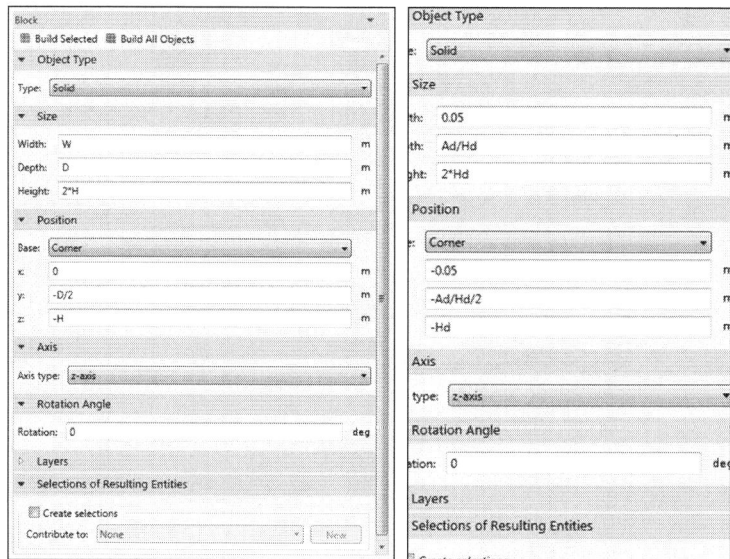

FIGURE 4.42: Data for geometry of the main and inlet spaces.

7. Now we build two blocks to cut the existing built ones, in order to create final computational domain using the symmetry. Create Block3, and in the Block window, under the Size section, enter 4 for all Width, Depth, and Height. Under the Position section, enter -1, -2, -4 for *x*, *y*, and *z*, respectively. The Base selection should read Corner. In order to perform the Boolean cut operation, click on Difference in the Geometry tab. In the Difference window, add uni1 to the list for Objects to add section, and blk3 to the Objects to subtract section. Users may need to activate the on/off buttons. Clear the Keep interior boundaries check box and click on the Build Selected icon. Similarly, create Block4 and in the Block window, under Size section enter 3, 2, 5 for all Width, Depth, and Height, respectively. Under Position section, enter -0.2, -2, -1 for *x*, *y*, and *z*, respectively. The Base selection should read Corner. Click on Difference in the Geometry tab. In the Difference window, add dif1 to the list for Objects to add section, and blk4 to the Objects to subtract section. Clear the Keep interior boundaries check box and click on the Build All Objects icon.

8. Now we build the geometry for the warm air inlet and ventilation outlet. Create two more blocks (Block5 and Block6) and enter their dimensions, as shown in Figure 4.43. Click on the Build All Objects button.

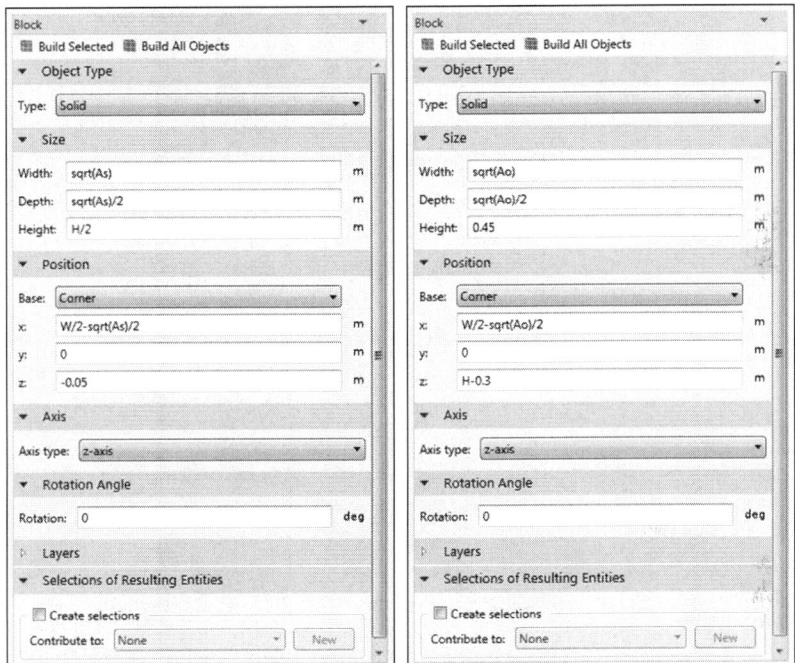

FIGURE 4.43: Data for the dimensions of warm air inlet and ventilation outlet.

For quality meshing purposes we make use of the Virtual Operations facility. This type of operation merges the adjacent domains in the model geometry but keep them, as virtual surfaces/edges, for meshing. It is useful to assign the sections of the ventilation inlet and exhaust outlet which are inside the room for meshing control purposes.

9. In the Geometry tab, click on Virtual Operations and select Mesh Control Domains. In the corresponding window, select and add domains 2 and 5 (the sections located inside the room domain) to the Input list. Click on the Build All icon. The Final geometry is shown in Figure 4.44.

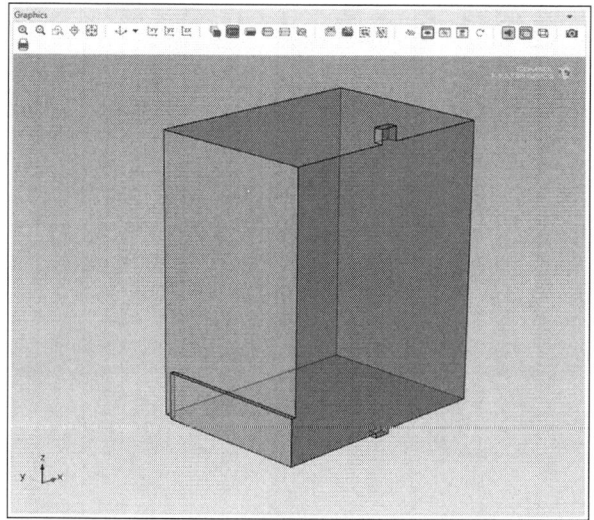

FIGURE 4.44: Geometry of the ventilated room with warm air inlet, exhaust, and fresh air inlet.

10. To add material properties, From the Home tab, click on Materials and select Add Material. Click on Built-In>Air and then click on the Add to Component button. Since we have heated air moving in the room, we keep the density as a function of temperature, but for now consider the viscosity as a constant. Click on the Fluid 1 node under the Non-Isothermal Flow (nitf) in the Model Builder window. In the Fluid window, locate Dynamic viscosity section and select User defined and enter 5e-4. Close the Add Material window.

Users may notice that the viscosity of air (at room temperature, 300K) is about 1.983e-5 Pa.s, which is about 25 times less than what we entered for the constant viscosity. This method is used, and recommended by the COMSOL model document, for calculating an initial guess/solution before calculating the final solution to help the convergence of the final solution, especially when the Reynolds number is relatively low.

Now we add the buoyancy force per unit volume.

11. Click on the Physics tab and from the ribbon click on Domains and select Volume Force. In the corresponding set-

tings window, add domain 1 to the Selection list. Locate the
Volume Force section and enter -g_const*nitf.rho for the
z-component of F.

12. Now we define the fluid flow boundary conditions. For warm
air inlet, from the Physics tab click on Boundaries and select
Inlet. In the Inlet window, select and add inlet boundary
13 to the Selection list. Locate Boundary conditions section
and enter 0.13 for Turbulent intensity (since a high level of
turbulence coming in with the warm air jet flow) and 0.01[m]
for Turbulent length scale. In the Velocity section enter Us
for Normal inflow velocity. Similarly for the fresh air inlet,
create another Inlet and select boundary 1 and velocity Ud
for it and leave other entries as default. We now set the outlet
boundary condition. From the Physics tab click on Boundaries and select Outlet. In the Outlet window, select and add
exhaust boundary 10 to the Selection list. Locate the Pressure Conditions section and select the Normal flow check
box.

13. Now we define the heat transfer boundary conditions. For
thermal condition at the warm air inlet, click on the Physics
tab and from Boundaries select Temperature. In the Temperature settings window select, and add inlet boundary 13
to the Selection list. Locate the Temperature section and
enter Tsource for T_0 field. Similarly, for the fresh air inlet
create another Temperature boundary and assign boundary
1 to its Selection list and Tdiff for its Temperature. We also
define the convective heat flux condition for the walls. From
the Physics tab, click on Boundaries and select Heat Flux. In
the Heat Flux settings window add wall boundaries 6, 8, 17
(i.e. all side walls except the symmetry plane) to the Selection
list. Locate the Heat Flux section and select the Inward heat
flux check box and enter 0.4 for h and Tout for T_{ext}. For the
exhaust outlet thermal condition, on the Physics toolbar, click
Boundaries and choose Outflow. In the Outflow settings window, add boundary 10 of the exhaust to the Selection list.

14. Symmetry is now defined. On the Physics toolbar, click
Boundaries and choose Symmetry, Flow. In the settings
window, add boundary 2 (the symmetry plane) to the Selec-

tion list. Similarly create a Symmetry, Heat boundary and assign boundary 2 to it.

15. We set the initial condition for the pressure. Click on the Initial Values node in the Model Builder window. In the Initial Values settings window, locate the Initial Values section and enter 1.2[kg/m^3]*9.81[m/s^2]*(H+0.15[m]-z) for Pressure p.

The model physics setting is complete. Now we recommend building two meshes for this model. Mesh 1 is an automatic physics-controlled and relatively a finer mesh, and Mesh 2 is a user-defined mesh based on the instructions given by the COMSOL model manual. We use Mesh 2 for this example and explain why a relatively finer mesh (like Mesh 1) is required for further study.

16. Click on the Mesh 1 node in the Model tree, and in the Mesh settings window select Fine for Element size. Click on Build All. It takes few minutes to build this mesh, which consists of more than 800 thousand elements. A close up of Mesh 1 near the vent exhaust and the symmetry wall is shown in Figure 4.45.

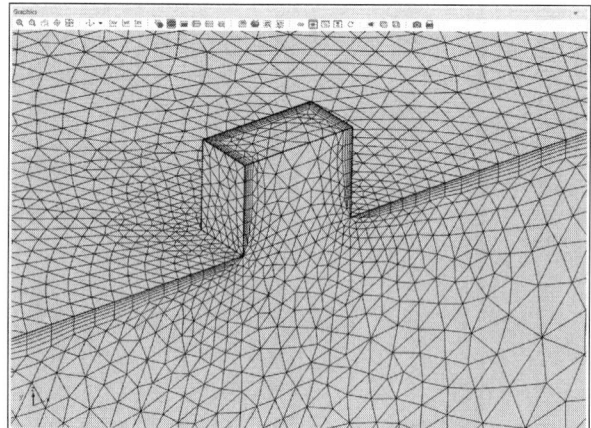

FIGURE 4.45: Mesh 1 close-up near the ventilation exhaust outlet.

17. Click on the Mesh tab and from the ribbon bar click on Add Mesh. The Mesh 2 node appears in the model tree. In the Mesh settings window, select Fine for Element size. Right-

click on the Mesh 2 node and select Edit Physics-Induced Sequence. Click on the Size 1 node. In the Size settings window, clear all boundaries selection from the list, located under the Geometric Entity Selection section. Select and add boundaries 4, 5, 9, 11, 12, 14, 15, 16 to the Selection list. These are boundaries on the sides of the both inlets (i.e. 4, 5, 12, 14, 15) and exhaust (i.e. 9, 11, 16) to the room. Locate the Element Size section and select the Custom check box. Locate the Element Size Parameters section and select the Maximum element size check box and edit the field by entering 0.015. Now we set the parameters for meshing the warm air inlet and ventilation exhaust domains. Click on Mesh 2>Free Tethrahedral 1>Size1 (if Size 1 node does not exist, create one). In the Size window select domains 2, 3, 4, 5. Locate the Element Size section and select the Custom check box. Under the Element Size Parameters section, select the Maximum element growth rate check box and edit the field by typing in 1.05. Finally we set the parameters for boundary layer mesh. Click on Mesh 2>Boundary Layer 1>Boundary Layer Properties1. In the Boundary Layer Properties window, locate the Boundary Layer Properties section and edit the Number of boundary layers field to 4, and Thickness adjustment factor to 3. Click on the Build All button. It takes few minutes to build this mesh, which consists of more than 600 thousand elements. A close-up of Mesh 2 near the ventilation exhaust is shown in Figure 4.46.

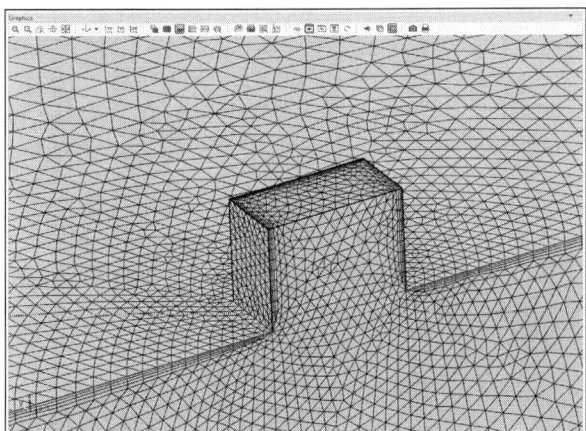

FIGURE 4.46: Mesh 2 close-up near the ventilation exhaust outlet.

18. Right-click on the Study1 node in the Model Builder window and rename it Study 1-Constant viscosity mu=5e-4[Pa*s]. This is optional, but useful to easily identify this Study against another one that we will build later for this model. Since we have two types of meshes (i.e. Mesh 1 and Mesh 2), selection should be made for running the model using both in sequence. Click on the Step 1: Stationary node and in the corresponding window locate the Mesh Selection section. Under the Mesh list, users can choose between Mesh 1 or Mesh 2. Select Mesh 2. Recall Mesh 2 is a user-defined fine resolution mesh.

19. The model is ready to run. It is useful, but optional, to cancel the automatic update of plotting the results during computation. Click on the Results node and in the corresponding window clear Automatic update of the plots check box. From the Home tab, click on Compute. It takes about 5 hours for computations to finish on a typical computer. Therefore, a workstation-level computer power is highly recommended for practical applications.

20. Default results may not appear in the Graphics window, since automatic updates of the plots was turned off. Click on the Results node in the model tree, and check the box for Automatic update of plots. In order to have a measure of the mesh resolution adequacy, it is useful to plot the cell Reynolds number. This is defined based on element sizes and varies through the flow domain. If it is too large then the mesh is coarse. Click on the Slice 1 node and in the Slice settings window replace Expression with nitf.cellRe. The main flow domain has a maximum cell Reynolds number close to 18, which indicates that mesh resolution is fine enough. Figure 4.47 shows the result.

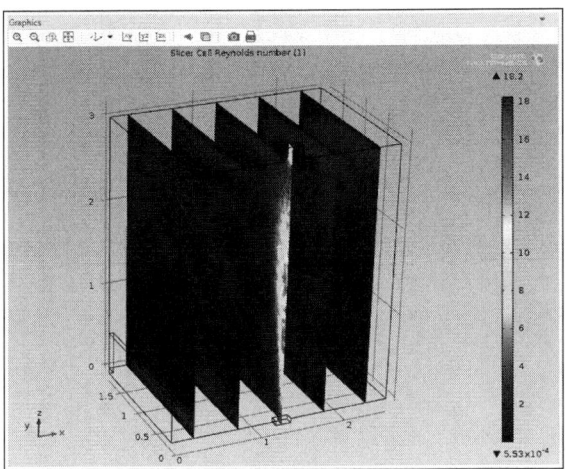

FIGURE 4.47: Cell Reynolds number for constant viscosity fluid flow.

Having the solution for constant viscosity, we now solve the model using a temperature dependent viscosity.

21. To set the viscosity of the fluid, click on the Fluid 1 node, under Non-Isothermal Flow (*nitf*). In the Fluid settings window locate the Dynamic viscosity section and select From material, for μ.

22. We add another study case for running the model. Click on the Study tab and then on the Add Study button, in the Study ribbon group. In the Add Study window, select Stationary for the list under Studies and click on the Add Study icon. Click on the Add Study button in the ribbon. A new Study 2 node appears in the model tree. Rename this to Study 2-Viscosity from material. To use the existing solution (i.e. Study 1) as the initial condition, click on Step 1: Stationary node, located under Study 2-Viscosity from material, and in the Stationary settings window locate the Values of Dependent Variables section. Check the box for Initial values of variables solved for, and select Solution, for Method and Study 1-Constant viscosity mu=5e-4[Pa*s], for Study.

23. The model is now ready for computation. Click on the Compute button in the Study toolbar.

24. Default results appear in the Graphics window. We plot the streamlines inside the room. Right-click on Temperature

(nitf)1 > Surface 1 and select Disable. Right-click on Temperature (nitf)1 and select Streamline. In the Streamline settings window locate the Data section and choose Solution 2 from the Data set list. Locate the Streamline Positioning section and choose Uniform density from the Positioning list and enter 0.07 for Separating distance. Locate the Coloring and Style section and choose Ribbon for Line type. To set the streamline coloring scale as the temperature, right-click on Temperature (nitf) 1 node and select Color Expression. In the Color Expression window expand the Range section and check the Manual color range box. Type in 293 and 300 for Minimum and Maximum, respectively. Locate the Coloring and Style section and choose the Thermal from Color table list. Click the Plot button. Click on the Transparency button in the Graphics toolbar. The result is shown in Figure 4.48.

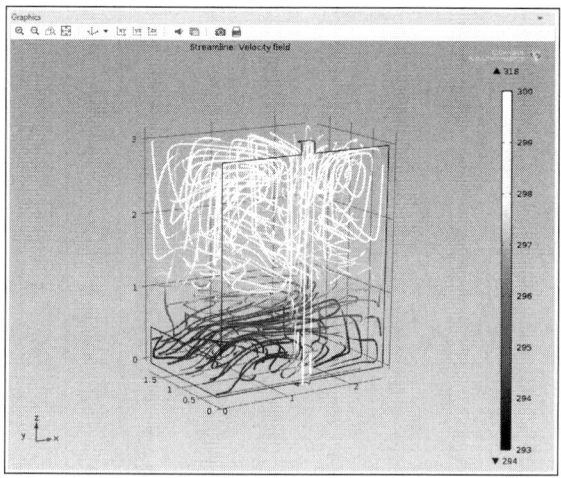

FIGURE 4.48: Streamlines visualizing the velocity field, colored by the temperature scale.

25. Now we draw some isothermal surfaces to show the stratified air layers in the room. Click on the Results tab in the toolbar and then select 3D Plot Group from the ribbon, in the Plot Group list. In the 3D Plot Group settings window locate the Data section and choose Solutions 2 from Data set list. From the 3D Plot Group 5 ribbon choose Isosurface. In the Isosurface settings window, locate the Expression section and choose degC from Unit list. Locate Levels section and choose Levels from the Entry method list and type in 22, 23, 24, 25, 26 for Levels. In the Coloring and Style section,

choose ThermalLight from the Color table list. Click on the
Plot button. The result is shown in Figure 4.49.

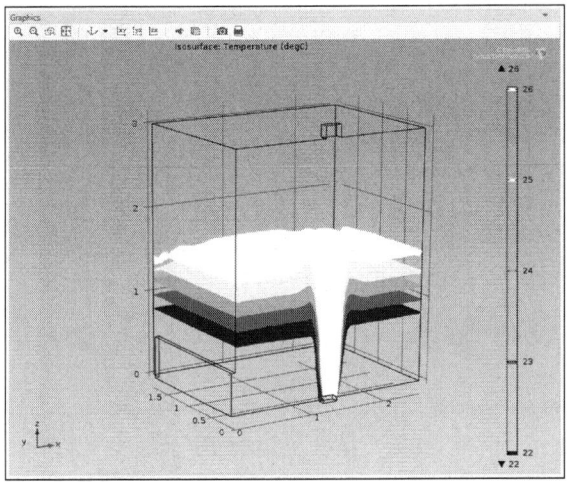

FIGURE 4.49: Isosurfaces of the temperature visualizing stratified air layers.

To validate the model against experimental data [94] we graph
the temperature through the center line of the inlet into the
room and extending up to the exhaust. The experimental data
are depicted in Table 4.4.

TABLE 4.4: Experimental data for temperatures measured
along the room height from inlet [94].

Z (m)	T (°C)	Z(m)	T (°C)
0.1073	44.5917	0.2042	43.5027
0.2732	41.8693	0.3285	40.4174
0.3698	38.9655	0.425	37.3321
0.4941	36.0617	0.5494	34.6098
0.6323	33.1579	0.7153	31.8875
0.826	30.98	0.9229	30.0726
1.0336	28.9837	1.1444	28.4392
1.2551	27.7132	1.3659	27.5318
1.5045	27.1688	1.6291	26.8058
1.7538	26.6243	1.8923	26.4428
2.0171	26.4428	2.1417	26.0799
2.2803	26.0799	2.4189	26.0799
2.5574	25.8984	2.6405	25.8984

26. First we plot a graph using the experimental data. Click on the Results>Tables>Table1 node in the model tree. In the Table settings window click on the Import button. Locate and import the file 'displacement_ventilation_exp.txt'. This file is available on the accompanying disk, or users can build a*.txt format file using the data provided in Table 4.4. The values will appear in the Table 1 window. Click on the Table Graph icon in the Table window toolbar. A new node 1D Plot Group 6>Table Graph 1 will appear in the model tree. Click on the Table Graph 1 node. In the Table Graph settings window locate the Data section and choose Column 2 for *x*-axis data. Locate the Coloring and Style section and choose None for Line, and Circle for Marker, and In data points for Positioning. Expand the Legends section and check the Show legends box. Choose Manual for legends and enter Exp. Data, under Legends. Click Plot.

27. Now we extract data for a line starting from the center of the inlet and ending at the exhaust. Right-click Data Sets and select Cut Line 3D from the list. In the Cut Line 3D settings window, choose Solutions 2 for Data set and enter W/2 for Point 1 *x*: and W/2 and H for Point 2 x: and *z*:, respectively.

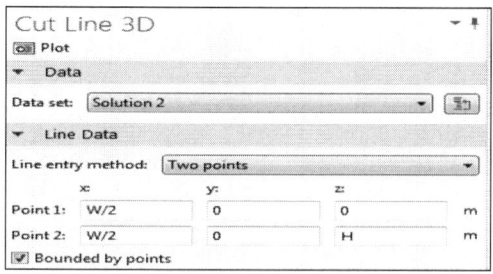

28. To draw the solution on the cut line, right-click on the 3D Plot Group 5 node, in the model tree, and select Line Graph. In the Line Graph settings window, locate *y*-Axis Data section and enter *z*. Locate the *x*-Axis data section and choose Expression from the list for Parameter. Enter T for Expression and degC for Unit. Locate the Coloring and Style section and enter 3 for Width. Expand the Legends section and check Show legends box. Choose Manual for legends and enter Model under Legends. Click on 1D Plot Group 6 and

locate the Plot Settings section. Check the boxes and enter T[degC] for x-axis label and z[m] for y-axis label. Locate the Axis section and check the box for Manual axis limits. Enter 0 and 46 for x minimum and x maximum, respectively, and 0 and 3 for y minimum and y maximum, respectively. Click the Plot button. The final results are shown in Figure 4.50. The graph shows that the experimental data drops faster along the height distance in the room compared to modeling results. The discrepancies in temperatures are a few degrees, and the cause could be the insulation of the wall and/or radiation effects [94] that cools the room faster as moving up in the room. For further investigation we could use a finer mesh (may be Mesh 1) for existing model or another turbulence model, like low-Re k-ε. This is left as exercise problems for users.

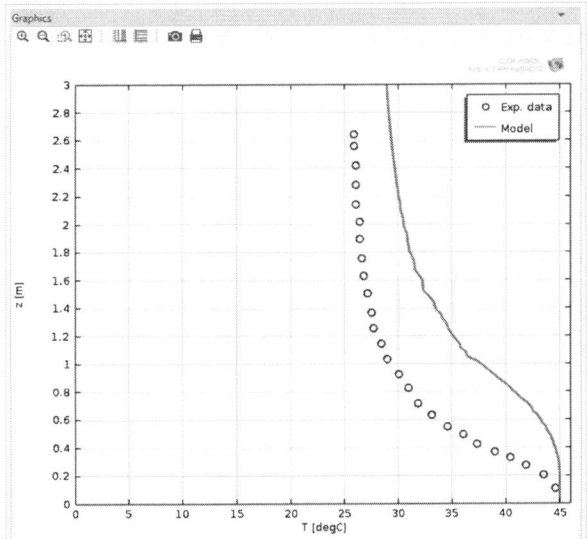

FIGURE 4.50: Experimental data and modeling results for temperatures along a line in the center of the room.

Example 4.5: Modeling of turbulent flow for a jet in cross-flow

For this example, we model the flow of a jet stream which is injected into another main stream flow. Usually the jet enters the main stream flow with higher momentum due to its higher average velocity relative to that of the main stream. This type of flow occurs in many industrial and natural applications, for instance during landing and take-off of airplanes, oil refinery piping, waste discharge into reservoirs or atmosphere. In any of these flow types turbulence occurs and enhances mixing or, in general, the exchange of mass and energy between the jet and the mainstream flows. Karvinen and Ahlstedt [95] modeled the experimental work of Özcan and Larsen [96] done for a similar settings. Former authors used several turbulence models (i.e. k-ε, versions of k-ω, versions of low-Re k- ε, and Reynolds stress) for the purpose of comparison and validation with those experimental results of the latter. We use the above mentioned references for validation of our modeling results, using the k-ε model, for this example. The main channel and jet flow domains (with a diameter of 24 mm) are shown in Figure 4.51, schematically.

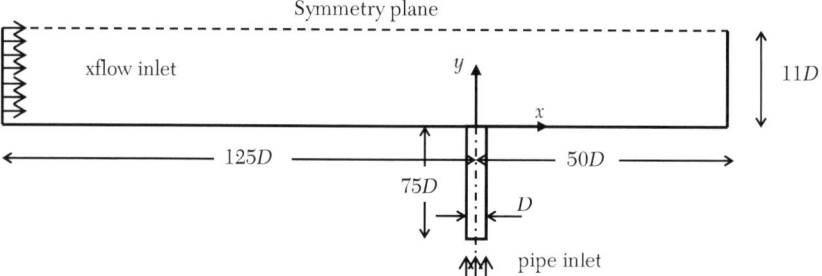

FIGURE 4.51: Schematic of cross-flow and jet flows geometry (not-to-scale). All dimensions are given as multiples of pipe diameter, D = 24 mm.

The center of coordinate system is located at the center-line of the pipe, $x=0$ located at the bottom of the channel. The mainstream cross-flow enters from left into the domain and jet flow enters through the pipe moving upwards (i.e. in positive y-direction).

Solution

    1. Launch COMSOL and from File>Save as, in the New window, and save the model as Example 4.5. Click on the Model Wizard icon.

    2. From the Select Space Dimension window, click on the 2D icon. The Select Physics window will appear. Select Fluid Flow>Single-Phase Flow>Turbulent Flow>Turbulent Flow, k-ε (spf). Click on Add icon.

    3. Click on the Study icon/arrow, and from the Select Study window, under Preset Studies, click on Stationary. Click on the Done icon. The COMSOL Desktop interface will appear. In the Geometry settings window locate the Units section and change the Length unit to mm.

4. Now we will make a list of input data used for this model as parameters. From the Home tab ribbon toolbar, click on Parameters. In the Parameters window, enter the data (case sensitive) as shown in Figure 4.52. Alternatively, this data can be imported from the accompanying disk.

Name	Expression	Value	Description
D	24[mm]	0.024000 m	pipe diameter
Fluid_dens	1.225[kg/m^3]	1.2250 kg/m³	air density
Fluid_vis	1.7894e-5[kg/m/s]*fact	1.7894E-5 kg/(...	air dyn. viscosity
Uinx	1.435[m/s]	1.4350 m/s	x-flow bulk inlet velocity
Iinx	2.5e-2	0.025000	x-flow turbulece intensity
Uinp	4.95[m/s]	4.9500 m/s	jet bulk inlet velocity
Uinf	1.5[m/s]	1.5000 m/s	free stream velocity
fact	1	1.0000	viscosity muliplier
Iinp	5.2e-2	0.052000	jet inlet turbulece intensity

FIGURE 4.52: Parameters data for Example 4.5.

5. To draw the geometry, click on the Geometry tab and choose Rectangle. Draw an arbitrary triangle in the Graphics window. Click on the Rectangle 1 (*r1*) node, in the model tree located under Geometry 1, and in the Rectangle settings window locate the Size section and enter 125*D+50*D for Width and 11*D for Height. Locate the Position section and enter -125*D for x and 0 for y. Create Rectangle 2 (*r2*), and in its corresponding settings window, enter *D* for Width and 75*D for Height, enter -D/2 for x and -75*D for y. Click on the Build All Objects button. Users may have to click on Zoom Extents, in the Graphics window toolbar, to see the whole built geometry.

6. For meshing purposes we create two lines which extend the pipe boundaries into the cross-flow domain. From the Geometry toolbar, click on Draw Line and draw an arbitrary line in the Graphics window (right-click to release the cursor). In the Bézier Polygon settings window locate the Polygon Segments section and click on Segment 1 (linear).

In the Control points section enter -D/2 for x and 0 for y for point 1, and -D/2 for *x* and 11*D for y for point 2. Click Build Selected. Similarly create another line by right-clicking on the *Bézier* Polygon 1 (*b*1) node, in the model tree, and choose Duplicate. Modify both x coordinates to read D/2. Click Build Selected. In order to use these lines, just created for meshing purposes, click on the Virtual Operations button, located in the ribbon toolbar under the Geometry tab, and choose Mesh Control Edges. In the Mesh Control Edges select and add the lines (6 and 10) to the list under Edges to include. Click on Build Selected. Notice that the two lines disappear from the geometry in the Graphics window.

7. To add material properties, Click on Turbulent Flow, *k-ε* (spf)>Fluid Properties 1. In the Fluid Properties window locate the Fluid Properties section and select User defined for both Density and Dynamic viscosity, and enter Fluid_dens and Fluid_vis, respectively (note that these entries should match those given in Parameters, i.e. case-sensitive).

8. Now we apply the boundary conditions. Click on the Physics tab and expand Boundaries from the toolbar, then select Symmetry. In the Symmetry settings window select and add upper boundary (boundary 3) to the Selection list. Similarly, from Boundaries, choose Outlet, and in the Outlet settings window, add boundary 9 to the Selection list and check both boxes for Normal flow and Suppress backflow, located in the Pressure Conditions section. We have two air inlets, one for the pipe entrance and the other for the cross-flow entrance. Click on Boundaries from the toolbar and select Inlet. Right-click on the Inlet 1 node that appears in the model tree and rename it Inlet 1-pipe. In the Inlet settings window add pipe entrance boundary (boundary 5) to the Selection list. Locate Boundary condition section and enter Iinp for Turbulence intensity and 0.07*D/2 for Turbulence length scale. In the Velocity section enter Uinp for U_0. Similarly, create another inlet boundary for cross-flow. Click on Boundaries from the toolbar and select Inlet. When the Inlet 2 node appears in the model tree, right-click on it, and rename it Inlet 2-xflow. In the Inlet settings window add cross-flow entrance bound-

ary (boundary 1) to the Selection list. Locate the Boundary condition section and enter Iinx for Turbulence intensity and 0.0057 for Turbulence length scale. In the Velocity section enter Uinx for U_0. We also modify the initial condition, to help the convergence of the solution. Click on Initial Values 1 node, in the model tree. In the Initial Values window locate Initial Values section and enter 0.1[bar] for p. The model physics setting is complete.

9. Now we create a mesh. For turbulence models (such as k-ε), which use a Wall function, and for the flow in this example, there exists boundary layer separation and maybe re-attachment in the domain of the cross stream. The mesh resolution is very important in order to capture the high-gradient quantities close to the wall. To have a control on mesh resolution we use a structured-mesh type for this model, using mesh parameters available for Mapped tool. Right-click on the Mesh 1 node and choose Mapped. In the Mapped settings window locate the Domain Selection section and choose Domain for Geometric entity level. Locate the Control section and un-check Smooth in the removed control entities check box. In the Advanced Settings section check box for Adjust evenly distributed edge mesh. Now we create a series of mesh controls for boundaries, using the Distribution tool. Right-click on the Mapped 1 node and choose Distribution. In the Distribution window select boundaries 1, 9, 12, and 13 (i.e. vertical lines across the main stream) and add them to the Selection list. Locate the Distribution section and select Predefined distribution type from the list under Distribution properties. Enter 132 for Number of elements and 10 for Element ratio. For the Distribution method, choose Geometric sequence from the list. Similarly, create four more Distribution nodes. The corresponding parameters are given in Table 4.5. To build the elements, click on the Mapped 1 node and add desired domains for meshing to the list and click on Build Selected. This option is useful to examine the elements and their resolution by building the mesh for each domain, separately. Alternatively, users may choose All domains, from Mapped settings window, and build the mesh for all domains at once, a total of 33,280 elements. A close-up of the resulted mesh near the jet area is shown in Figure 4.53.

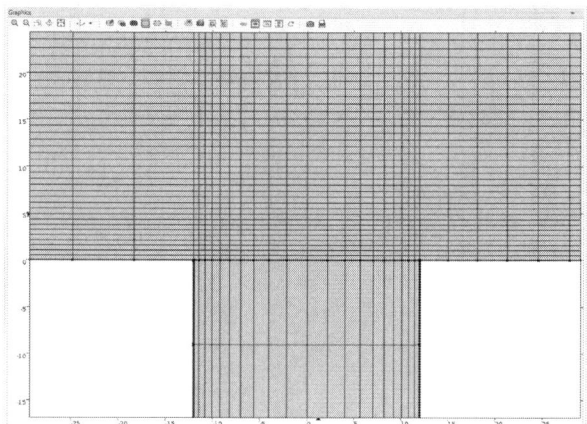

FIGURE 4.53: A close-up of structured mesh built near the jet area.

TABLE 4.5: Mesh control parameters for boundaries, using Distribution tool.

Mesh control	Boundaries symbols	Number of elements	Element ratio	Attributes
Distribution 1	B1, B2, B3, B4	132	10	Predefined distribution, Geometric sequence
Distribution 2	B5, B6	120	10	Predefined distribution, Geometric sequence
Distribution 3	B7, B8, B9	20	4	Predefined distribution, Geometric sequence, Symmetric
Distribution 4	B10, B11	100	10	Predefined distribution, Geometric sequence, Reverse direction
Distribution 5	B12, B13	80	5	Predefined distribution, Geometric sequence Reverse direction

```
              B6                    B9              B11
    ┌───────────────────┐                 ┌───────────────────┐
 B1 │                    B6        B2 │ B3        B11           │ B4
    └───────────────────┘                 └───────────────────┘
              B5                    B8              B10
                              B12   B13
                              └─────┘
                                B7
```

10. To help with the convergence, we gradually increase air velocity at the inlet of the pipe. Click on Study 1> Step 1:

Stationary node in the model tree. In the Stationary settings window, expand the Study Extensions section. Check box for Auxiliary sweep and add (click plus sign) Uinp (jet bulk inlet velocity) to the list under Auxiliary parameters. Enter 0.5, 2, 4.95 for Parameter value list.

11. Click on the Study tab and select Compute from the toolbar.

12. Default results for velocity will appear in the Graphics window, after computation is finished. Among the default results is the dimensionless wall distance. This value should not be very large relative to 11.06, as mentioned in the COMSOL manual. Click on the Wall Resolution (spf) node and in the 2D Plot Group settings window locate the Data section and choose 4.95 for Parameter value (Uinp). Click Plot. The Graphics window will show a value of 11.1 for dimensionless wall distance, which is quite acceptable. The mesh quality however, could be increased by manipulating the mesh Distribution controls to have a lower value for wall dimensionless distances, as well. We leave this as an exercise problem for users. Default velocity and pressure contours are shown in Figure 4.54.

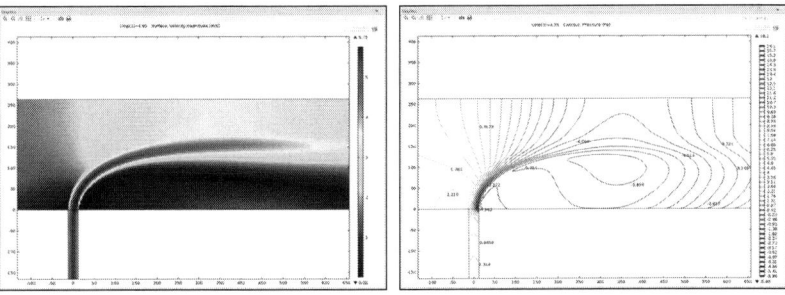

FIGURE 4.54: Velocity magnitude (left) and pressure contours (right) close to jet injection area.

13. For comparison and validation of the results we extract data from the solution at cross-sections located in the cross-flow at $\frac{x}{D} = 1.5, 1.83, 3, 3.67$. We are interested in variations of velocity components, turbulent kinetic energy, and turbulent kinetic energy dissipation values at these cross sectional locations. Click on the Results tab and choose Cut Line

2D from the ribbon toolbar. In the Cut Line 2D settings window, locate the Line Data section and choose Point and direction from the list for Line entry method. For Point enter 1.5*D for x and 0 for y. For Direction enter 0 for x and 1 for y. Rename Cut Line 2D 1 to Cut Line 2D 1-at x/D =1.5. Create three more cut-lines with x = 1.83*D, x = 3*D, x = 3.67*D; remaining parameters remain the same as Cut Line 2D 1. It is easier to use the Duplicate tool. Right-click on Cut Line 2D 1- at x/D = 1.5 node, in the model tree, and choose Duplicate from the list. In the corresponding settings window edit x *for* 1.83*D, and similarly create remaining cut-lines.

14. To draw the variables at the cut-lines, click on the Results tab and choose 1D Plot Group, from the ribbon toolbar. In the toolbar for 1D Plot Group 4, choose Line Graph. Rename Line Graph 1 to Line Graph 1- at x/D = 1.5. In the Line Graph settings window locate the Data section and choose Cut Line 2D 1- at x/D = 1.5 for Data set, and Last for Parameter selection (Uinp). Locate the y-Axis Data section and enter y/D for Expression. In the x-Axis data section, choose Expression for Parameter and enter u/Uinf under Expression. In the Coloring and Style section, enter 3 for Width. In the Legends section check the box for Show legends and enter *x/D* = 1.5 for Legends, after choosing Manual for Legends. Click Plot. Users may have to adjust the coordinate's axis scales. Similarly, create more line graphs and choose associated with the three remaining cut-lines. Results are shown in Figure 4.55. These results are comparable with those of Karvinen et al. [95]. The trend of variations, mainly movement of velocity peak upwards and to the right, is consistent with similar results reported by the references mentioned above.

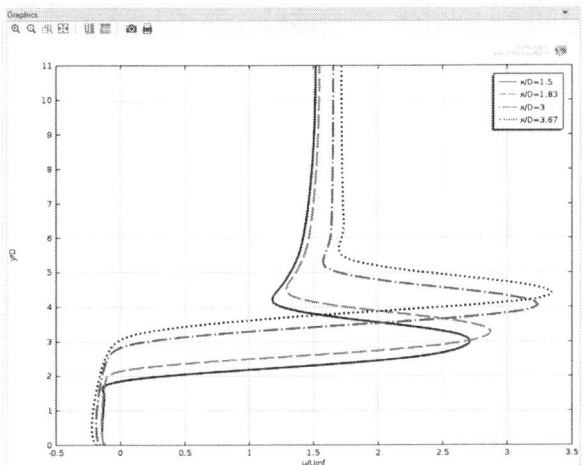

FIGURE 4.55: Dimensionless velocity component u/U_∞, at locations downstream of the jet.

15. For making plots for remaining variables of interest (i.e. v, k, and ε) simply use a duplicate of the plot group made for velocity component u. Right-click on the 1D Plot Group4-u profile and choose Duplicate; a new Plot group appears. Rename this plot group 1D Plot Group 4-v profile and in the corresponding Line Graph's change the Expression, under x-Axis Data to v/Uinf for all four Line Graphs. Similarly repeat the Duplicate process for k and modify the Expression under x-Axis Data to k/Uinf^2 for all corresponding four Line Graphs. Finally, repeat the Duplicate process for ε and modify the Expression, under x-Axis Data to ep/(Uinf^3/D) for all corresponding four Line Graphs. Results are shown in Figures 4.56–4.58.

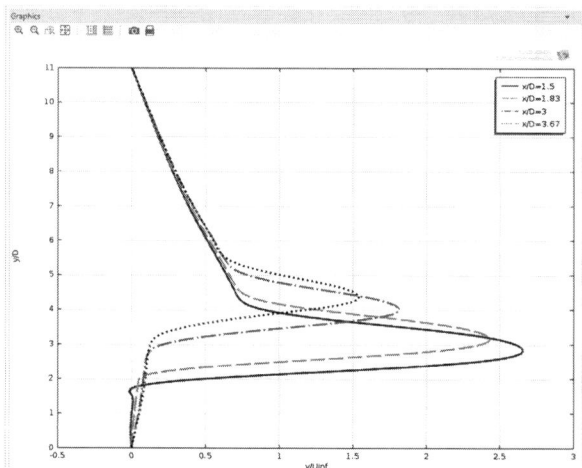

FIGURE 4.56: Dimensionless velocity component, at locations downstream of the jet.

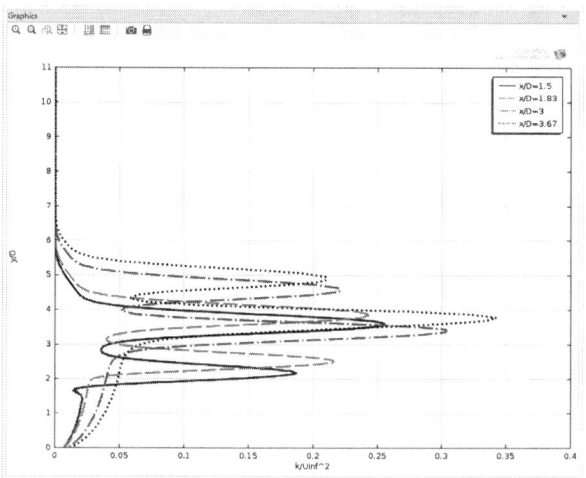

FIGURE 4.57: Dimensionless turbulent kinetic energy k / U_∞^2, at locations downstream of the jet.

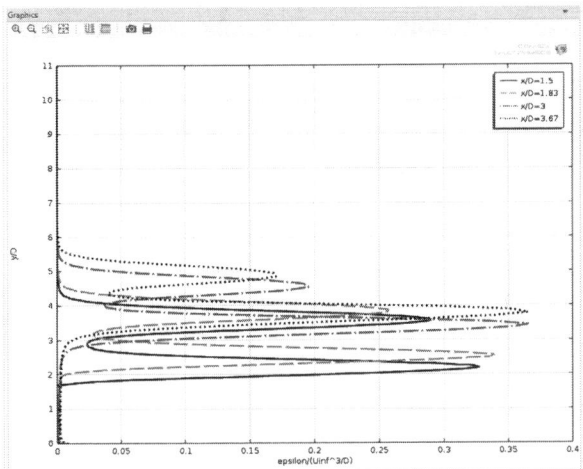

FIGURE 4.58: Dimensionless turbulent kinetic energy

dissipation $\left(\dfrac{\varepsilon D}{U_\infty^3}\right)$ at locations downstream of the jet.

16. We draw the streamlines close to the jet injection area. Click on the Results tab and choose the 2D Plot Group. From the 2D Plot Group choose Streamline. In the Streamline settings window, locate the Streamline Positioning section and select Start point controlled from the Positioning list and enter 30 for Points. Select Cut Line 2D- at x/D=10 (users should create this cut-line or use previously built ones, results vary accordingly) from the list for Along line. In the Coloring and Style section, adjust the parameters to get the desired settings. Right-click on the Streamline 1 node in the model tree, and select Color Expression. Results are shown in Figure 4.59. This figure clearly shows the recirculating zone, downstream of the jet injection location.

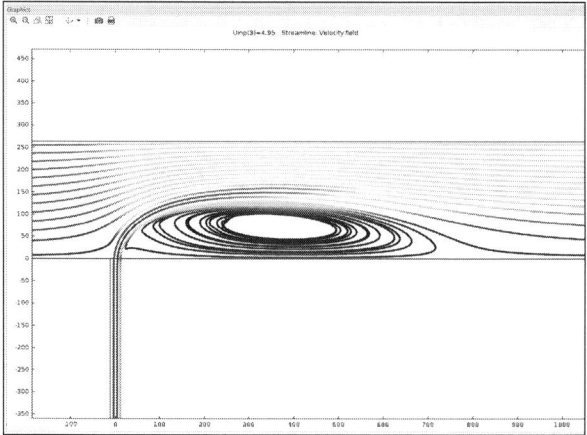

FIGURE 4.59: Streamlines close to the jet injection area.

Example 4.6: Modeling of turbulent flow over a circuit board in a duct

For this example, we model turbulent air flow inside a duct of an electronic package. The physical domain consists of a rectangular duct and protruding electronic components attached to the bottom of the duct, as shown in Figure 4.60. We use the L-VEL model for isothermal turbulent flow. Users may want to solve this example using the yPlus model and compare the results.

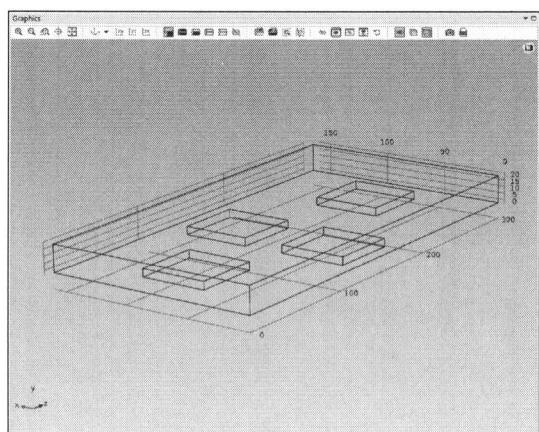

FIGURE 4.60: Geometry and dimensions of the duct with electronic heaters

As mentioned in Chapter 2, the governing equations for the turbulent models, used for this example, are those of RANS in addition to algebraic equations using an enhanced eddy viscosity model based on the local wall distance. In COMSOL, the physics interface therefore includes a wall distance equation. We use COMSOL 5 for building this model.

Solution:

1. Launch COMSOL and from File>Save as, in the New window, save the model as Example 4.6-flow in an IC duct-L-VEL. Click on the Model Wizard icon.

2. From the Select Space Dimension window, click on the 3D icon. The Select Physics window will appear. Select Fluid Flow>Single-Phase Flow>Turbulent Flow>Turbulent Flow, L-VEL (spf). Then click on the Add icon.

3. Click on the Study icon/arrow, and from the Select Study window, under Preset Studies, click on Stationary with Initialization. Click on Done icon. The COMSOL Desktop interface will appear. In the Geometry settings window locate the Units section and change the Length unit to mm.

4. Now we make a list of input data used as parameters for this models. From the Model tab ribbon toolbar, click on Parameters. In the Parameters window, enter the data (case sensitive) as shown in Figure 4.61. Alternatively, this data can be imported from the accompanying disc media. The data are also provided for modeling non-isothermal and/or conjugate heat transfer, covered as exercise problems.

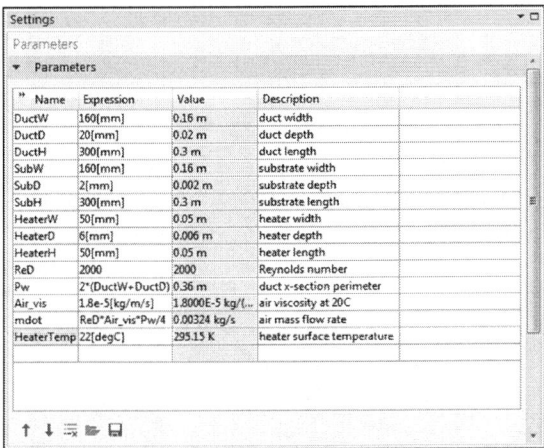

Name	Expression	Value	Description
DuctW	160[mm]	0.16 m	duct width
DuctD	20[mm]	0.02 m	duct depth
DuctH	300[mm]	0.3 m	duct length
SubW	160[mm]	0.16 m	substrate width
SubD	2[mm]	0.002 m	substrate depth
SubH	300[mm]	0.3 m	substrate length
HeaterW	50[mm]	0.05 m	heater width
HeaterD	6[mm]	0.006 m	heater depth
HeaterH	50[mm]	0.05 m	heater length
ReD	2000	2000	Reynolds number
Pw	2*(DuctW+DuctD)	0.36 m	duct x-section perimeter
Air_vis	1.8e-5[kg/m/s]	1.8000E-5 kg/(...	air viscosity at 20C
mdot	ReD*Air_vis*Pw/4	0.00324 kg/s	air mass flow rate
HeaterTemp	22[degC]	295.15 K	heater surface temperature

FIGURE 4.61: Parameters data for Example 4.6.

5. Now we will build model geometry. Click on Block, under the Geometry tab toolbar. In the Settings window for Block, enter DuctW, DuctD, DuctH for Width, Depth, and Height, respectively. In the Position section, select Corner for Base. Rename the Block 1(blk1) node, in the Model Builder window, to Duct. Similarly, build more blocks using the data provided in Table 4.6 and click on the Build All Objects icon. Reorient the geometry built in the Graphics window and click on Wireframe Rendering to see all objects. Final results are shown in Figure 4.60.

TABLE 4.6: Dimensions and data for building geometry Blocks in Example 4.6.

	Size			Position			
	Width	**Depth**	**Height**	**x**	**y**	**z**	**Base**
Block 1 (Duct)	DuctW	DuctD	DuctH	0	0	0	Corner
Block 2 (Heater1)	HeaterW	HeaterD	HeaterH	(DuctW-HeaterW)/2	0	25	Corner
Block 3 (Heater2)	HeaterW	HeaterD	HeaterH	(DuctW-HeaterW)/2	0	DuctH-25-HeaterH	Corner
Block 4 (Heater3)	HeaterW	HeaterD	HeaterH	40	0	DuctH/2	Center
Block 5 (Heater4)	HeaterW	HeaterD	HeaterH	DuctW-40	0	DuctH/2	Center

6. We will now remove/subtract the Heater1-4 from the flow domain. Click on Booleans and Partitions from the toolbar under the Geometry tab. Select Difference and in the Settings window add Duct (i.e. blk1) to the selection list for Objects to add. Click on the Active button in the Objects to subtract section and add Heater1-4 (i.e. blk1-4) to the. Click Build All Objects icon. We also make use of the geometry/flow symmetry and use half of the domain for modeling. We cut the domain in the middle at a plane parallel to y-z plane. Create another Block and rename it Symmetry block. In the Settings window, for this Block, Enter DuctW/2, 10*DuctD, DuctH for Width, Depth, and Height, respectively. Locate Position section, select Corner for Base, and enter -10*DuctD/2 for y. Click Build Selected. Now click on Booleans and Partitions from the toolbar under the Geometry tab. Select Difference, and in the Settings window add flow domain (i.e. dif2) to the selection list for Objects to add.

Click on the Active button in the Objects to subtract section, and add Symmetry block (i.e. blk6) to the list. Click the Build All Objects icon. Half of the model geometry will appear in the Graphics window.

7. To add fluid materials to the model, click on the Materials tab in the toolbar, and click on Add Material. Type in Air, and click the Search button. Expand Built-In and click on Air, then click on the +Add to component button. Click on the Add Material button in the toolbar to close the Materials window. Air is automatically added to the flow domain.

8. Now we define the boundary conditions for the turbulent flow physics. Click on the Physics tab in the toolbar, and select Inlet from the list under Boundaries. In the Settings for Inlet window, add boundary 3 to the selection list. Locate the Boundary Condition section and select Mass flow from the list. Enter mdot for Mass flow rate. Similarly, create an Outlet and Symmetry boundaries and assign boundaries 8 and 1 to them, respectively. All other boundaries have no-slip boundary conditions, by default.

9. To create finite elements for the flow domain, click on the Mesh 1 node in the Model Builder window. In the Settings window, select Extra coarse for Element size. Click Build All. A view of created mesh is shown in Figure 4.62.

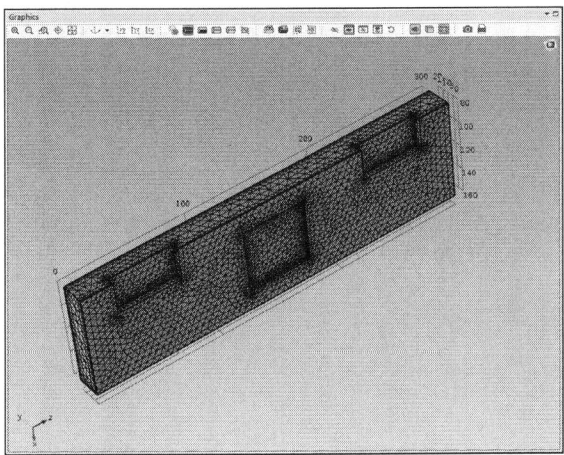

FIGURE 4.62: Mesh created for turbulent flow domain in Example 4.6.

10. The model is ready to run. Click on the Study tab from the toolbar, and click on the Compute button. Default results, for velocity and pressure, appear in the graphics window, when computations are finished. Users may watch the convergence curve, shown in the Convergence Plot 1 window, while computations are in progress. Users may run the model with a finer mesh, however, Wall Resolution is close to 4.5. See the COMSOL Manual for detail. Velocity at slice surfaces and pressure contours are shown in Figure 4.63, after some modifications to the default results.

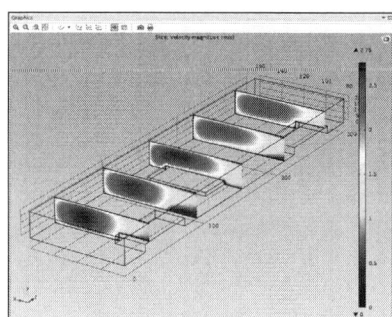 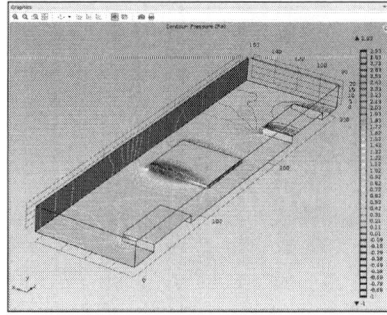

FIGURE 4.63: Velocity surfaces at slice planes (left) and pressure contours (right), ReD=2000.

11. We now run the model for a series of values of Reynolds number. Click on the Study tab, in the toolbar, and select Parametric Sweep. In the Settings window, locate the Study Settings section and add ReD (Reynolds number) to the list under Parameter name. Enter 2000, 4000, 6000 in the space under Parameter value list. Click on the Compute button. The results for ReD = 6000 are shown in Figure 4.64.

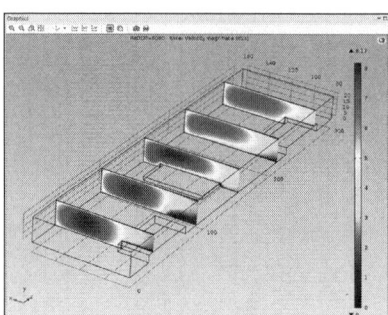 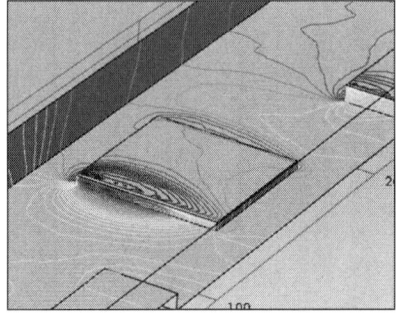

FIGURE 4.64: Velocity surfaces at slice planes (left) and pressure contours (right), ReD=6000.

12. Build an App for this model using Application Builder. Users may follow similar instructions given for Example 3.1. A built App is shown in Figure 4.65.

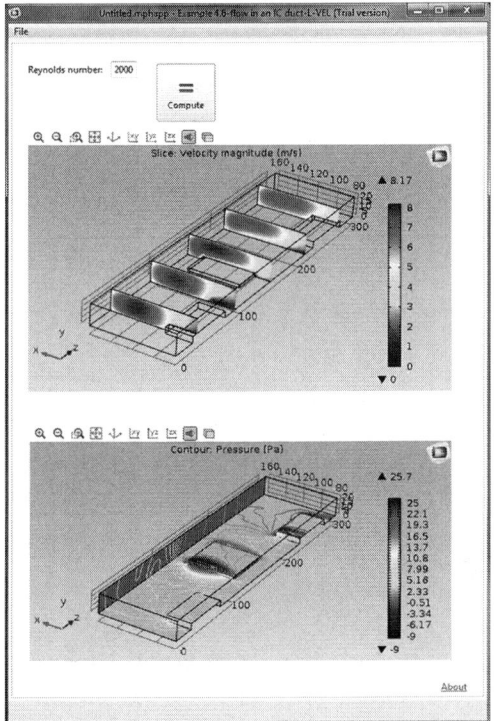

FIGURE 4.65: A built App for model Example 4.6.

EXERCISES

4.1. Model Buice 2D asymmetric diffuser, as mentioned in Example 4.1, using S-A turbulent model. Compare the results with the solution for the SST model.

4.2. Discuss and extract dimensionless distance from the wall to the cell center, for model Example 4.1.

4.3. Draw the curves for pressure coefficient for Example 4.1.

4.4. Repeat model in Example 4.1, by using a finer mesh and a Mapped mesh. Compare the results with the solution provided.

4.5. Build an application for Example 4.1, using the COMSOL 5 Application Builder.

4.6. Repeat model in Example 4.2 for another user-defined mesh. Use the Mapped mesh tool for building the mesh. Investigate the results for pressure, drag and lift coefficients versus existing experimental results. Users may benefit from the guidelines given in the COMSOL Model Libraries and by following the instructions given for naca0021 airfoil and NASA Quest at *http://quest.nasa. gov/people/bios/aero/fjournals/duque/grids.html*.

4.7. In model Example 4.2, investigate the Wall Resolution value. This variable is the dimensionless distance from the wall to the nearest element/cell center. In the COMSOL manual [55] (i.e. variable l_c^*) it is recommended to have its value close to unity. Compare the results of exercise problem 4.5 against model Example 4.2.

4.8. Repeat model Example 4.2 using the SST turbulence model. Perform mesh sensitivity analysis and compare the results against Example 4.2.

4.9. Build an application for Example 4.2, using the COMSOL 5 Application Builder.

4.10. Repeat model Example 4.3 using the $k\text{-}\varepsilon$ model and compare your results with this model example, as well as of Homicz [92].

4.11. Modify model Example 4.3 by building a new mesh with user-defined parameters. As a reference you may want to use CFD_Module/Single-Phase_Benchmarks/pipe_elbow available from the COMSOL web site (to registered users). Solve the example with the new mesh and compare the results.

4.12. Repeat model Example 4.3 by increasing the bulk velocity to higher values than 5 m/s, hence higher Reynolds number. Investigate the effect of the Reynolds number on the flow through the bend pipe, for example, a graph of Dean number versus Reynolds number, etc.

4.13. Build an application for Example 4.3, using the COMSOL 5 Application Builder.

4.14. Repeat model Example 4.4 using a relatively finer mesh, like Mesh 1 as mentioned for this example.

4.15. Repeat model Example 4.4 using a different turbulent model, like $k\text{-}\varepsilon$ and compare your results with the existing one.

4.16. Build an application for Example 4.4, using the COMSOL 5 Application Builder.

4.17. Repeat model Example 4.5 using a finer mesh and compare the results, specifically for turbulent kinetic energy and its dissipation, as well as mesh statistics and dimensionless wall distance.

4.18. Repeat model Example 4.5 using the SST turbulent model and compare the results.

4.19. Repeat model Example 4.5, having jet as salt water entering fresh water mainstream.

4.20. Repeat model Example 4.5, assuming the jet fluid is hot water, perhaps at a temperature of 85°C, and main stream is a river, perhaps at a temperature of 10°C. Plot the temperature variations downstream of the jet and discuss your results.

4.21. Build an application for Example 4.5, using the COMSOL 5 Application Builder.

4.22. Repeat Example 4.6 using the yPlus turbulent model and compare the results.

4.23. Repeat Example 4.6 by adding a substrate to the electronic heaters and use non-isothermal flow with conjugate heat conduction in the substrate.

4.24. Build an application for Example 4.6, using the COMSOL 5 Application Builder.

APPENDIX

DERIVATION OF GOVERNING EQUATIONS

This appendix is divided into several sections and is very much involved with derivations of the main turbulence governing equations, as presented in the previous chapters. We use tensor notation with Einstein's summation convention, [41] for presenting these equations and their derivations. We start with deriving some relations pertinent to averaging operation, focusing on time averaging, used for Reynolds decomposition of turbulent quantities. After that we continue with derivations of the exact forms of the governing equations for RANS, Reynolds stresses, turbulent kinetic energy, and turbulent kinetic energy dissipation rate.

AVERAGING RELATIONS

For performing time-averaging operation on the Reynolds-decomposed terms involved in N-S equations, we require some relations handy and ready to use. We derive these relations in this section. We cover only time averaging of variables involved, but the method could be readily used for other averaging methods, like ensemble averaging [50]. Without losing generality, we define

$\overline{\Phi}(x) = \lim\limits_{T \to \infty} \dfrac{1}{T} \int_0^T \Phi(x,t)\,dt.$ In this section, we further define the the operator $\mathbb{L}$ for the purpose of abbreviation.

$$\mathbb{L}(\cdots) \equiv \lim\limits_{T \to \infty} \dfrac{1}{T} \int_0^T (\cdots)\,dt$$

Therefore, we have $\mathbb{L}(\Phi) = \overline{\Phi}$. Preliminary averaging relations for two arbitrary functions Φ and ψ can easily be derived from this definition:

$$\overline{\Phi + \psi} = \overline{\Phi} + \overline{\psi}, \quad \overline{\overline{\Phi}\psi} = \overline{\overline{\Phi}\psi}, \quad \overline{\overline{\Phi}} = \overline{\Phi}, \quad \overline{\dfrac{\partial \Phi}{\partial t}} = \dfrac{\partial \overline{\Phi}}{\partial t}.$$

Using the averaging definition, we derive averaging relations for Reynolds-decomposed velocity vector $u_i = \overline{u}_i + u'_i$ as given by equation (A1):

$$\overline{u}_i = \mathbb{L}(u_i) = \mathbb{L}(\overline{u}_i + u'_i) = \overline{\overline{u}}_i + \overline{u'}_i = \overline{u}_i + \overline{u'}_i \tag{A1}$$

From equation (A1) we conclude that $\overline{\overline{u}}_i = \overline{u}_i$ and $\overline{u'}_i = 0$. Now we derive the averaged expression for a product term, such as $u_i u_j$ (a tensor of 2nd rank):

$$\overline{u_i u_j} = \mathbb{L}(u_i u_j) = \mathbb{L}\left\{ (\overline{u}_i + u'_i)(\overline{u}_j + u'_j) \right\}$$

$$= \mathbb{L}(\overline{u}_i \overline{u}_j) + \mathbb{L}\left(u'_i u'_j \right) + \mathbb{L}(\overline{u}_i u'_j) + \mathbb{L}(\overline{u}_j u'_i)$$

$$= \overline{\overline{u}_i \overline{u}_j} + \overline{u'_i u'_j} + \overline{\overline{u}_i u'_j} + \overline{\overline{u}_j u'_i}$$

$$= \overline{u}_i \overline{u}_j + \overline{u'_i u'_j} + \underbrace{\overline{u}_i \overline{u'_j}}_{=0} + \underbrace{\overline{u}_j \overline{u'_i}}_{=0}$$

The last two terms are zero, since $\overline{u'}_j = \overline{u'}_i = 0$ (see Equation A1). Finally we have:

$$\overline{u_i u_j} = \overline{u}_i \overline{u}_j + \overline{u'_i u'_j} \tag{A2}$$

We also derive the average relation for a term such as $u_{i,j} u_j$, or

$$\overline{u_{i,j} u_j} = \mathbb{L}\left(u_{i,j} u_j \right) = \mathbb{L}\left\{ (\overline{u}_i + u'_i)_{,j} (\overline{u}_j + u'_j) \right\}$$

$$= \mathbb{L}\left(\overline{u}_{i,j} \overline{u}_j \right) + \mathbb{L}\left(u'_{i,j} u'_j \right) + \mathbb{L}\left(\overline{u}_{i,j} u'_j \right) + \mathbb{L}\left(\overline{u}_j u'_{i,j} \right)$$

$$= \overline{\overline{u}_{i,j}\overline{u}_j} + \overline{u'_j u'_{i,j}} + \overline{\overline{u}_{i,j}u'_j} + \overline{\overline{u}_j u'_{i,j}}$$

$$= \overline{u}_{i,j}\overline{u}_j + \overline{u'_j u'_{i,j}} + \underbrace{\overline{u}_{i,j}\overline{u'}_j}_{=0} + \underbrace{\overline{u}_j \overline{u'}_{i,j}}_{=0}$$

Therefore, we have

$$\overline{u_j u_{i,j}} = \overline{u}_j \overline{u}_{i,j} + \overline{u'_j u'_{i,j}} \tag{A3}$$

The next relation is a triple correlation, or averaging a term like $u_i u_j u_k$ (a 3rd rank tensor):

$$\overline{u_i u_j u_k} = \mathbb{L}\left(u_i u_j u_k \right) = \mathbb{L}\left\{ \, (\overline{u}_i + u'_i)(\overline{u}_j + u'_j)(\overline{u}_k + u'_k) \right\}$$

After working out the products and averaging each term, we get

$$\overline{u_i u_j u_k} = \overline{\overline{u}_i \overline{u}_j \overline{u}_k} + \overline{\overline{u}_i u'_j u'_k} + \overline{\overline{u}_j u'_i u'_k} + \overline{\overline{u}_k u'_i u'_j} + \overline{u'_i u'_j u'_k}$$

$$+ \underbrace{\overline{\overline{u}_i \overline{u}_k u'_j}}_{=0} + \underbrace{\overline{\overline{u}_j \overline{u}_k u'_i}}_{=0} + \underbrace{\overline{\overline{u}_i \overline{u}_j u'_k}}_{=0}$$

The last three terms are zero, since $\overline{u'_j} = \overline{u'_i} = \overline{u'_k} = 0$, and finally , we get

$$\overline{u_i u_j u_k} = \overline{u}_i \overline{u}_j \overline{u}_k + \overline{u}_i \overline{u'_j u'_k} + \overline{u}_j \overline{u'_i u'_k} + \overline{u}_k \overline{u'_i u'_j} + \overline{u'_i u'_j u'_k} \tag{A4}$$

Similarly, averaging relations could be derived for higher rank tensors or expressions, if required.

REYNOLDS AVERAGED NAVIER-STOKES EQUATIONS (RANS)

In order to derive RANS equations, we start from N-S equations for incompressible fluids. Without losing generality, similar derivations are applicable for N-S equations for compressible fluids. We repeat N-S equations (2.4–2.5) here, for convenience:

$$\rho\left(\frac{\partial u_i}{\partial t} + u_j u_{i,j} \right) = -p_{,i} + \mu u_{i,jj} \tag{2.4}$$

$$u_{i,i} = 0 \tag{2.5}$$

After substituting $u_i = \overline{u}_i + u'_i$, $p = \overline{p} + p'$, and $\rho \cong \overline{\rho}$ we get, from equation (2.5):

$u_{i,i} = (\bar{u}_i + u'_i)_{,i} = \bar{u}_{i,i} + u'_{i,i} = 0,$ and after averaging $\overline{\bar{u}_{i,i}} = \bar{\bar{u}}_{i,i} + \underbrace{(\overline{u'_i})_{,i}}_{=0} = \bar{u}_{i,i} = 0.$

Hence both mean velocity (concluded from the latter relation) and velocity fluctuation (concluded from the former relation) satisfy continuity:

$$\bar{u}_{i,i} = 0$$
$$u'_{i,i} = 0 \qquad (A5)$$

Similarly, from equation (2.4), we get

$$\rho\left(\frac{\partial(\bar{u}_i + u'_i)}{\partial t} + (\bar{u}_j + u'_j)(\bar{u}_i + u'_i)_{,j}\right) = -(\bar{p} + p')_{,i} + \mu(\bar{u}_i + u'_i)_{,jj} \quad (A6)$$

or

$$\rho\frac{\partial(\bar{u}_i)}{\partial t} + \rho\frac{\partial(u'_i)}{\partial t} + \rho\left(\bar{u}_j\bar{u}_{i,j} + \bar{u}_j u'_{i,j} + u'_j\bar{u}_{i,j} + u'_j u'_{i,j}\right)$$
$$= -\bar{p}_{,i} - p'_{,i} + \mu(\bar{u}_{i,jj} + u'_{i,jj})$$

After averaging, using averaging relations derived in the previous section, we get

$$\rho\frac{\partial(\bar{u}_i)}{\partial t} + \rho\frac{\partial(\overset{=0}{\overline{u'_i}})}{\partial t} + \rho\left(\bar{u}_j\bar{u}_{i,j} + \overset{=0}{\overline{\bar{u}_j u'_{i,j}}} + \overset{=0}{\overline{u'_j\bar{u}_{i,j}}} + \overline{u'_j u'_{i,j}}\right)$$

$$= -\bar{p}_{,i} - \overset{=0}{\overline{p'}}_{,i} + \mu(\bar{u}_{i,jj} + \overset{=0}{\overline{u'}}_{i,jj})$$

Or after collecting the non-zero terms, we get

$$\rho\frac{\partial(\bar{u}_i)}{\partial t} + \rho\bar{u}_j\bar{u}_{i,j} + \underbrace{\rho\overline{u'_j u'_{i,j}}}_{new\,term} = -\bar{p}_{,i} + \mu\bar{u}_{i,jj}$$

The new term, on the L.H.S could be written as

$\overline{\rho u'_j u'_{i,j}} = \rho(\overline{u'_i u'_j})_{,j} - \rho\underbrace{\overline{u'_i u'_{j,j}}}_{=0} = \rho(\overline{u'_i u'_j})_{,j},$ after back-substituting for this term and moving it to the R.H.S, we get the

$$\rho\frac{\partial(\bar{u}_i)}{\partial t} + \rho\bar{u}_j\bar{u}_{i,j} = -\bar{p}_{,i} + \mu\bar{u}_{i,jj} - \rho(\overline{u'_i u'_j})_{,j} \qquad (A7)$$

Where Reynolds stress is defined as $R_{ij} = -\rho(\overline{u'_i u'_j})$. Using Boussinesq's hypothesis $R_{ij} = \mu_T\left(\bar{u}_{i,j} + \bar{u}_{j,i}\right) - \frac{2}{3}\rho k\delta_{ij}$, after substituting into equation (A7) for these relations, we get

$$\rho \frac{\partial(\overline{u}_i)}{\partial t} + \rho \overline{u}_j \overline{u}_{i,j} = -\overline{p}_{,i} + \mu \overline{u}_{i,jj} + \mu_T \left(\overline{u}_{i,jj} + \underbrace{\overline{u}_{j,ij}}_{\substack{=(u_{j,j}),i \\ =0}} \right) - \frac{2}{3} \rho(k\delta_{ij})_{,j}$$

The last term on the R.H.S. could be written as $-\frac{2}{3}\rho(k\delta_{ij})_{,j} =$

$-\frac{2}{3}\rho(k_{,j})\delta_{ij} = -\frac{2}{3}\rho(k_{,i})$ (since $\delta_{ij} = 1$ for $i = j$, otherwise is zero). We

combine this term with the pressure, then $-\left(\overline{p} + \frac{2}{3}\rho k\right)_{,i} = -\overline{p}*_{,i}$.

After substituting these terms back into (A7), we get

$$\rho \frac{\partial(\overline{u}_i)}{\partial t} + \rho \overline{u}_j \overline{u}_{i,j} = -\overline{p}*_{,i} + (\mu + \mu_T)\overline{u}_{i,jj} \qquad (A8)$$

Equation (A8) along with $\overline{u}_{i,i} = 0$ are RANS equations with Boussinesq's hypothesis included and form a system of equations which are determinate (see Chapter 2).

EXACT EQUATION FOR REYNOLDS STRESS TRANSPORT

We start from momentum equation for velocity u_i, which could be interpreted as momentum per unit mass. The conservation form of momentum equation, for an incompressible fluid, is

$$\rho \left(\frac{\partial u_i}{\partial t} + (u_i u_k)_{,k} \right) = -p_{,i} + \sigma_{ik,k} \qquad (A9)$$

where $\sigma_{ik} = \mu(u_{i,k} + u_{k,i})$, is the deviatoric part of the stress tensor. Now, we multiply both sides by u_j and get

$$\rho \left(u_j \frac{\partial u_i}{\partial t} + u_j(u_i u_k)_{,k} \right) = -u_j p_{,i} + u_j \sigma_{ik,k} \qquad (A10)$$

Similarly, we can start with the momentum equation for velocity u_j, and then multiply it by u_i, to get

$$\rho \left(u_i \frac{\partial u_j}{\partial t} + u_i(u_j u_k)_{,k} \right) = -u_i p_{,j} + u_i \sigma_{jk,k} \qquad (A11)$$

Now we sum up equations (A10) and (A11), to get

$$\rho\left(u_j\frac{\partial u_i}{\partial t} + u_i\frac{\partial u_j}{\partial t}\right) + \rho\left(u_j(u_iu_k)_{,k} + u_i(u_ju_k)_{,k}\right) \tag{A12}$$

$$= -\left(u_jp_{,i} + u_ip_{,j}\right) + u_j\sigma_{ik,k} + u_i\sigma_{jk,k}$$

The goal is to manipulate some of the terms in equation (A12), such that we get an equation for the transport of u_iu_j. The second term, on the L.H.S. could be written as

$$\rho(u_j(u_iu_k)_{,k} + u_i(u_ju_k)_{,k}) = \rho(u_iu_ju_k)_{,k}\text{, since}$$

$$\rho(u_iu_ju_k)_{,k} = \rho\left[u_j(u_iu_k)_{,k} + u_{j,k}(u_iu_k)\right] = \rho\left[u_j(u_iu_k)_{,k} + u_i\underbrace{\left(u_ku_{j,k}\right)}_{=(u_ku_j)_{,k}}\right],$$

where continuity $u_{k,k} = 0$, is used for deriving the last term in the bracket. Using this expression and back substitute it to equation (A12), after some manipulation, we get

$$\rho\left(\frac{\partial(u_iu_j)}{\partial t}\right) + \rho(u_iu_ju_k)_{,k} = -\left(u_jp_{,i} + u_ip_{,j}\right) + u_j\sigma_{ik,k} + u_i\sigma_{jk,k} \tag{A13}$$

Now we perform Reynolds decomposition operation on Eq. (A13) by substituting for $u_i = \overline{u}_i + u'_i$, $p = \overline{p} + p'$, and $\rho \cong \overline{\rho}$, to get

$$\rho\left(\frac{\partial(\overline{u}_i + u'_i)(\overline{u}_j + u'_j)}{\partial t}\right) + \rho\left[(\overline{u}_i + u'_i)(\overline{u}_j + u'_j)(\overline{u}_k + u'_k)\right]_{,k} \tag{A14}$$

$$= -(\overline{u}_j + u'_j)(\overline{p} + p')_{,i} - (\overline{u}_i + u'_i)(\overline{p} + p')_{,j}$$

$$+(\overline{u}_j + u'_j)\sigma_{ik,k} + (\overline{u}_i + u'_i)\sigma_{jk,k}$$

The deviatoric stress terms should be decomposed accordingly, for example:

$\sigma_{ik} = \mu\left(u_{i,k} + u_{k,i}\right) = \mu\left[\left(\overline{u}_{i,k} + \overline{u}_{k,i}\right) + \left(u'_{i,k} + u'_{k,i}\right)\right]$. We define then, the two parts of the Reynolds decomposed deviatoric stress, as

$$\overline{\sigma}_{ik} = \mu\left(\overline{u}_{i,k} + \overline{u}_{k,i}\right)$$

$$\sigma'_{ik} = \mu\left(u'_{i,k} + u'_{k,i}\right)$$

Therefore $\sigma_{ik} = \overline{\sigma}_{ik} + \sigma'_{ik}$, which by substitution into equation (A14), we get

$$\rho\left(\frac{\partial(\overline{u}_i + u'_i)(\overline{u}_j + u'_j)}{\partial t}\right) + \rho\left[(\overline{u}_i + u'_i)(\overline{u}_j + u'_j)(\overline{u}_k + u'_k)\right]_{,k}$$
$$= -(\overline{u}_j + u'_j)(\overline{p} + p')_{,i} - (\overline{u}_i + u'_i)(\overline{p} + p')_{,j} \tag{A15}$$
$$+(\overline{u}_j + u'_j)(\overline{\sigma}_{ik} + \sigma'_{ik})_{,k} + (\overline{u}_i + u'_i)(\overline{\sigma}_{jk} + \sigma'_{jk})_{,k}$$

Now we perform averaging operation on equation (A15), using the averaging relations (equations A1–A4) obtained in the previous section (including $\overline{u'}_j = \overline{u'}_i = 0$), to get

$$\rho\left(\frac{\partial(\overline{u}_i\overline{u}_j)}{\partial t} + \frac{\partial(\overline{u'_i u'_j})}{\partial t}\right) \tag{A16}$$
$$+\rho\left(\overline{u}_i\overline{u}_j\overline{u}_k + \overline{u}_i\overline{u'_j u'_k} + \overline{u}_j\overline{u'_i u'_k} + \overline{u}_k\overline{u'_i u'_j} + \overline{u'_i u'_j u'_k}\right)_{,k}$$
$$= -\overline{(\overline{u}_j + u'_j)(\overline{p} + p')}_{,i} - \overline{(\overline{u}_i + u'_i)(\overline{p} + p')}_{,j}$$
$$+(\overline{u}_j\overline{\sigma}_{ik,k} + \overline{u'_j\sigma'_{ik,k}}) + (\overline{u}_i\overline{\sigma}_{jk} + \overline{u'_i\sigma'_{jk,k}})$$

Notice that $\overline{\sigma}'_{ik} = \mu\left(\overline{u'}_{i,k} + \overline{u'}_{k,i}\right) = 0$ (since $\overline{u'_i} = \overline{u'_k} = 0$). The terms on the R.H.S, which involve pressures, are derived further. We manipulate the first term, and equivalently apply the results to the second term, as well.

$\overline{(\overline{u}_j + u'_j)(\overline{p} + p')}_{,i} = \overline{u}_j\overline{p}_{,i} + \overline{u'_j p'}_{,i} + \underbrace{\overline{u}_j\overline{p'}_{,i}}_{=0} + \underbrace{\overline{u'_j}\overline{p}_{,i}}_{=0}$, but the second term, could be further written as

$$\overline{(\overline{u}_j + u'_j)(\overline{p} + p')}_{,i} = \overline{u}_j\overline{p}_{,i} + \overline{u'_j \underbrace{(p - \overline{p})}_{=p'}}_{,i} = \overline{u}_j\overline{p}_{,i} + \overline{u'_j p}_{,i} - \underbrace{\overline{u'_j \overline{p}}_{,i}}_{=0}.$$

Therefore, we have $\overline{(\overline{u}_j + u'_j)(\overline{p} + p')}_{,i} = \overline{u}_j\overline{p}_{,i} + \overline{u'_j p}_{,i}$ and similarly, $\overline{(\overline{u}_i + u'_i)(\overline{p} + p')}_{,j} = \overline{u}_i\overline{p}_{,j} + \overline{u'_i p}_{,j}$. After substituting these relations back into equation (A16), we get

$$\rho\left(\frac{\partial(\overline{u}_i\overline{u}_j)}{\partial t} + \frac{\partial(\overline{u'_i u'_j})}{\partial t}\right) \tag{A17}$$
$$+\rho\left(\overline{u}_i\overline{u}_j\overline{u}_k + \overline{u}_i\overline{u'_j u'_k} + \overline{u}_j\overline{u'_i u'_k} + \overline{u}_k\overline{u'_i u'_j} + \overline{u'_i u'_j u'_k}\right)_{,k}$$
$$= -\overline{u}_j\overline{p}_{,i} - \overline{u'_j p}_{,i} - \overline{u}_i\overline{p}_{,j} - \overline{u'_i p}_{,j} + (\overline{u}_j\overline{\sigma}_{ik,k} + \overline{u'_j\sigma'_{ik,k}})$$
$$+(\overline{u}_i\overline{\sigma}_{jk} + \overline{u'_i\sigma'_{jk,k}})$$

Equation (A17) involves terms that are averaged velocity correlations, as well as their fluctuations correlations. Therefore if we find an equation which governs the averaged velocity correlation (or mean kinetic energy), and subtract it from (A17), we should get what we are aiming for, i.e., the transport equation for Reynolds stress. Now we derive the mean-kinetic energy equation.

Mean-kinetic Energy Equation

We start this derivation from equation (A7), where after re-arranging some of the terms, using continuity, and changing dummy index $j \to k$, we get [13]:

$$\rho \frac{\partial (\bar{u}_i)}{\partial t} + \rho \left(\bar{u}_i \bar{u}_k + \overline{u'_i u'_k} \right)_{,k} = -\bar{p}_{,i} + \mu \bar{u}_{i,kk}$$

The last term on the R.H.S could be written in terms of averaged deviatoric stress, since $\bar{\sigma}_{ik,k} = \mu \left(\bar{u}_{i,k} + \bar{u}_{k,i} \right)_{,k} = \mu (\bar{u}_{i,kk} + \underbrace{u_{k,ik}}_{=0})$, the

last term is zero due to continuity. Therefore we get

$$\rho \frac{\partial (\bar{u}_i)}{\partial t} + \rho \left(\bar{u}_i \bar{u}_k + \overline{u'_i u'_k} \right)_{,k} = -\bar{p}_{,i} + \bar{\sigma}_{ik,k} \qquad (A18)$$

We write equation (A18) for $\bar{u}_j$ (i.e., changing free index $i \to j$) to get

$$\rho \frac{\partial (\bar{u}_j)}{\partial t} + \rho \left(\bar{u}_j \bar{u}_k + \overline{u'_j u'_k} \right)_{,k} = -\bar{p}_{,j} + \bar{\sigma}_{jk,k} \qquad (A19)$$

Now perform a 'cross-operation' by algebraically multiplying equation (A18) by $\bar{u}_j$ and sum it up with equation (A19) when multiplied by $\bar{u}_i$. The result is shown with equation (A20), after some re-arrangements:

$$\rho \frac{\partial (\bar{u}_i \bar{u}_j)}{\partial t} + \rho \left(\bar{u}_i \bar{u}_j \bar{u}_k \right)_{,k} = -\bar{u}_i \bar{p}_{,j} - \bar{u}_j \bar{p}_{,i} + \bar{u}_j (\bar{\sigma}_{ik} - \rho \overline{u'_i u'_k})_{,k}$$
$$+ \bar{u}_i (\bar{\sigma}_{jk} - \rho \overline{u'_j u'_k})_{,k} \qquad (A20)$$

By performing a contraction operation (i.e. multiplying by δ_{ij}), we get the governing equation for mean-kinetic energy, as

$$\rho \frac{\partial (\bar{u}_i \bar{u}_i)}{\partial t} + \rho \left(\bar{u}_i \bar{u}_i \bar{u}_k \right)_{,k} = -2 \bar{u}_i \bar{p}_{,i} + 2 \bar{u}_i (\bar{\sigma}_{ik} - \rho \overline{u'_i u'_k})_{,k} \qquad (A21)$$

To arrive at the Reynolds stress transport equation, we subtract equation (A20) from equation (A17). The results is:

$$\rho \frac{\partial (\overline{u'_i u'_j})}{\partial t} + \rho \left(\bar{u}_k \overline{u'_i u'_j} + \overline{u'_i u'_j u'_k} \right)_{,k} \qquad (A22)$$
$$= -\overline{u'_j p_{,i}} - \overline{u'_i p_{,j}} + \overline{u'_j \sigma_{ik,k}} + \overline{u'_i \sigma_{jk,k}} - \rho \overline{u'_i u'_k} \bar{u}_{j,k} - \rho \overline{u'_j u'_k} \bar{u}_{i,k}$$

Substituting for $R_{ij} = -\rho(\overline{u_i'u_j'})$, we get

$$\frac{\partial R_{ij}}{\partial t} + \left(\overline{u}_k R_{ij}\right)_{,k} - \rho\left(\overline{u_i'u_j'u_k'}\right)_{,k} = \overline{u_j'p_{,i}} + \overline{u_i'p_{,j}} - \overline{u_j'\sigma_{ik,k}'} \qquad (A23)$$

$$-\overline{u_i'\sigma_{jk,k}'} - R_{ik}\overline{u}_{j,k} - R_{jk}\overline{u}_{i,k}$$

Equation (A23) is the exact equation governing Reynolds stresses transport. As discussed in Chapter 2, the third term on the L.H.S. is a new term and clearly shows that calculating Reynolds stress from N-S equations needs a *closure*.

EXACT EQUATION FOR TURBULENT KINETIC ENERGY TRANSPORT

To derive the turbulent kinetic energy equation we perform a contraction operation (i.e. multiply by δ_{ij}) on equation (A22), or

$$\rho\frac{\partial(\overline{u_i'u_i'})}{\partial t} + \rho\left(\overline{u}_k\overline{u_i'u_i'} + \overline{u_i'u_i'u_k'}\right)_{,k} = -2\overline{u_i'p_{,i}} + 2\overline{u_i'\sigma_{ik,k}'} \qquad (A24)$$

$$-2\rho\overline{u_i'u_k'}\overline{u}_{i,k}$$

After substituting for kinetic energy $k = \frac{1}{2}\overline{u_i'u_i'}$ (we use symbol k representing averaged turbulent kinetic energy, in this appendix, to avoid confusion with index k), we get

$$\rho\frac{\partial(k)}{\partial t} + \rho\left(\overline{u}_k k\right)_{,k} + \frac{1}{2}\rho\left(\overline{u_i'u_i'u_k'}\right)_{,k} = -\overline{u_i'p_{,i}} + \overline{u_i'\sigma_{ik,k}'} \qquad (A25)$$

$$-\rho\overline{u_i'u_k'}\,\overline{u}_{i,k}$$

We now manipulate, expand, and rearrange some of the terms in equation (A25).

$$\rho\left(\overline{u}_k k\right)_{,k} = \rho\left(\underbrace{\overline{u}_{k,k}}_{=0} k + \overline{u}_k k_{,k}\right) = \rho\overline{u}_k k_{,k}, \text{ and}$$

$$-\overline{u_i'p_{,i}} = -\overline{u_i'\left(\overline{p} + p'\right)_{,i}} = -\overline{u_i'}\,\overline{p}_{,i} - \overline{u_i'p_{,i}'} = -\underbrace{(\overline{u_i'p'})_{,i}}_{i \to k} = -(\overline{u_k'p'})_{,k}.$$

The last operation is legitimate, since i is a dummy index and could be changed to another index, such as k. Also, the second term on the

R.H.S. could be written as $\overline{u_i'\sigma_{ik,k}'} = \left(\overline{u_i'\sigma_{ik}'}\right)_{,k} - \overline{u_{i,k}'\sigma_{ik}'}$, which after substituting for $\sigma_{ik}' = \mu\left(u_{i,k}' + u_{k,i}'\right)$, we get

$$
\begin{aligned}
\left(\overline{u_i'\sigma_{ik}'}\right)_{,k} &= \mu\left(\overline{u_i'\left(u_{i,k}' + u_{k,i}'\right)}\right)_{,k} \\
&= \mu\left(\overline{u_{i,k}'u_{i,k}' + u_i'u_{i,kk}' + u_i'\underbrace{u_{k,ik}'}_{=0} + u_{i,k}'u_{k,i}'}\right) \\
&= \mu\left(\overline{\left(\tfrac{1}{2}u_i'u_i'\right)_{,kk}}\right) + \mu\overline{u_{i,k}'u_{k,i}'},
\end{aligned}
$$

and for $-\overline{u_{i,k}'\sigma_{ik}'} = -\mu\overline{u_{i,k}'\left(u_{i,k}' + u_{k,i}'\right)} = -\mu\overline{u_{i,k}'u_{i,k}'} - \mu\overline{u_{i,k}'u_{k,i}'}$. Therefore, finally we get (for the second term on the R.H.S. of [A25]);

$$
\overline{u_i'\sigma_{ik,k}'} = \mu\left(\overline{\left(\frac{1}{2}u_i'u_i'\right)_{,kk}}\right) - \mu\overline{u_{i,k}'u_{i,k}'} = \mu k_{,kk} - \mu\overline{u_{i,k}'u_{i,k}'}. \quad \text{After collect-}
$$

ing all terms and back-substituting into equation (A25), after some rearrangements, we get

$$
\rho\frac{\partial(k)}{\partial t} + \rho\bar{u}_k\,k_{,k} = -\rho\overline{u_i'u_k'}\,\bar{u}_{i,k} - \mu\overline{u_{i,k}'u_{i,k}'} + \mu k_{,kk} - \left(\frac{1}{2}\rho\overline{u_i'u_i'u_k'} + \overline{u_k'p'}\right)_{,k}
$$

This equation is a scalar equation, since k and p' are scalar quantities and all other terms have repeated indices (i.e. dummy indices). We change index $k \to j$ and finally, we get (after exposing the differentiations)

$$
\rho\frac{\partial k}{\partial t} + \rho\bar{u}_j\frac{\partial k}{\partial x_j} = -\rho\overline{u_i'u_j'}\frac{\partial\bar{u}_i}{\partial x_j} - \mu\overline{\frac{\partial u_i'}{\partial x_j}\frac{\partial u_i'}{\partial x_j}} \quad (A26)
$$

$$
+\mu\frac{\partial^2 k}{\partial x_j\partial x_j} - \frac{\partial}{\partial x_j}\left(\frac{1}{2}\rho\overline{u_i'u_i'u_j'} + \overline{p'u_j'}\right)
$$

Equation (A26) is the exact equation for transport of turbulent kinetic energy and recovers equation (2.17) from Chapter 2, after substituting for $-\rho\overline{u_i'u_j'}\dfrac{\partial\bar{u}_i}{\partial x_j} = R_{ij}\dfrac{\partial\bar{u}_i}{\partial x_j} = P_k$.

EXACT EQUATION FOR TURBULENT KINETIC ENERGY DISSIPATION TRANSPORT

The second term on the R.H.S. of equation (A26), after dividing by fluid density ρ, is $\nu \overline{\dfrac{\partial u_i'}{\partial x_j} \dfrac{\partial u_i'}{\partial x_j}}$. This term is actually the rate of kinetic energy dissipation per unit mass for a Newtonian incompressible fluid in isotropic turbulence flows [21]. Therefore, we define

$$\varepsilon = \nu \overline{\frac{\partial u_i'}{\partial x_j} \frac{\partial u_i'}{\partial x_j}} \tag{A27}$$

This term represents the amount of energy dissipated from turbulence and converted to heat (or internal energy) through viscosity. Its appearance in the kinetic energy transport equation (A26) motivates derivation of an equation which governs ε. Prior to deriving the exact equation governing for ε transport [13], we would like to mention that the second term on the R.H.S. of Eq. (A25) (i.e. $\overline{u_i' \sigma_{ik,k}'}$) contains total kinetic energy dissipation rate per unit mass. This can be shown as follow, after dividing by ρ:

$\left(\dfrac{1}{\rho}\right)\overline{u_i'\sigma_{ik,k}'} = \left(\dfrac{1}{\rho}\right)\overline{(u_i'\sigma_{ik}')_{,k}} - \left(\dfrac{1}{\rho}\right)\overline{u_{i,k}'\sigma_{ik}'}$. However, the last term can

be written as $\Upsilon = \left(\dfrac{1}{\rho}\right)\overline{u_{i,k}'\sigma_{ik}'} = \nu\overline{u_{i,k}'(u_{i,k}'+u_{k,i}')} = \nu\overline{u_{i,k}'u_{i,k}'} + \nu\overline{u_{i,k}'u_{k,i}'}$.

Using the definition for ε (i.e. Equation A27) we get $\Upsilon = \varepsilon + \nu\overline{u_{i,k}'u_{k,i}'}$. But last term can be further manipulated, using continuity, as

$$\Upsilon = \varepsilon + \nu\overline{(u_i'u_k')}_{,ik} \tag{A28}$$

Equation (A28) gives the total rate of dissipation for an incompressible fluid. The second term on the R.H.S. of equation (A28) vanishes for isotropic/homogeneous turbulence; hence we recover equation (A27). Also note that this term (i.e. the second term in Equation A28) is proportional to the second derivative of Reynolds stresses and when its value is zero, or negligible, we have $\Upsilon \cong \varepsilon$, [21].

To derive the ε-equation, we start with equation (A6) and then subtract equation (A7) from it. The result is

$$\rho\left(\frac{\partial u_i'}{\partial t} + \overline{u}_j u_{i,j}' + u_j'\overline{u}_{i,j} + u_j'u_{i,j}'\right) = -p_{,i}' + \mu u_{i,jj}' + \rho(u_i'u_j')_{,j} \tag{A29}$$

We perform three operations on equation (A29): first differentiate it (i.e. $\dfrac{\partial \circ}{\partial x_k} \equiv \circ_{,k}$), second multiply it by $u'_{i,k}$ (inner product), and third, average it. We write each term of equation (A29) separately while performing these operations on each, as

$$\rho\left(\overline{u'_{i,k}\left(\dfrac{\partial u'_i}{\partial t}\right)}_{,k}\right) = \rho\dfrac{\partial}{\partial t}\left(\dfrac{1}{2}\overline{(u'_{i,k}u'_{i,k})}\right),$$

$$\rho\left(\overline{u'_{i,k}\left(\overline{u}_j u'_{i,j}\right)_{,k}}\right) = \rho\left(\overline{u}_{j,k}\overline{u'_{i,j}u'_{i,k}} + \overline{u}_j\overline{u'_{i,k}u'_{i,kj}}\right)$$

$$= \rho\left(\overline{u}_{j,k}\overline{u'_{i,kj}u'_{i,j}} + \overline{u}_j\left(\dfrac{1}{2}\overline{u'_{i,k}u'_{i,k}}\right)_{,j}\right),$$

$$\rho\left(\overline{u'_{i,k}\left(u'_j\overline{u}_{i,j}\right)_{,k}}\right) = \rho\left(\overline{u'_{i,k}u'_{j,k}}\,\overline{u}_{i,j} + \overline{u'_{i,k}u'_j}\,\overline{u}_{i,jk}\right),$$

$$\rho\left(\overline{u'_{i,k}\left(u'_j u'_{i,j}\right)_{,k}}\right) = \rho\left(\overline{u'_{i,k}u'_{j,k}u'_{i,j}} + \overline{u'_{i,k}u'_j u'_{i,jk}}\right)$$

$$= \rho\left(\overline{u'_{i,k}u'_{j,k}u'_{i,j}} + \overline{u'_j\left(\dfrac{1}{2}u'_{i,k}u'_{i,k}\right)_{,j}}\right),$$

$$\overline{u'_{i,k}\left(-p'_{,ik}\right)} = -\overline{u'_{i,k}p'_{,ik}}, \text{ and}$$

$$\rho\left(\overline{u'_{i,k}\left(\overline{u'_i u'_j}\right)_{,jk}}\right) = \rho\underbrace{\overline{u'_{i,k}}}_{=0}\left(\left(\overline{u'_i u'_j}\right)_{,jk}\right) = 0$$

$$\mu\overline{u'_{i,k}\left(u'_{i,jj}\right)_{,k}} = \mu\left(\overline{u'_{i,k}u'_{i,jk}}\right)_{,j} - \mu\overline{u'_{i,jk}u'_{i,jk}}$$

$$= \dfrac{1}{2}\mu\left(\overline{u'_{i,k}u'_{i,k}}\right)_{,jj} - \mu\overline{u'_{i,jk}u'_{i,jk}}$$

Back-substituting all terms into equation. (A29), multiply both sides by 2ν, and let $\varepsilon = \nu\overline{u'_{i,k}u'_{i,k}}$. After rearrangement and exposing derivatives notations, we get

$$\rho\dfrac{\partial \varepsilon}{\partial t} + \rho\overline{u}_j\dfrac{\partial \varepsilon}{\partial x_j}$$

$$= \underbrace{-2\mu\overline{\dfrac{\partial u'_i}{\partial x_k}\dfrac{\partial u'_j}{\partial x_k}\dfrac{\partial \overline{u}_i}{\partial x_j}} - 2\mu\overline{\dfrac{\partial u'_i}{\partial x_k}\dfrac{\partial u'_i}{\partial x_j}\dfrac{\partial \overline{u}_j}{\partial x_k}} - 2\mu\overline{\dfrac{\partial u'_i}{\partial x_k}u'_j}\dfrac{\partial^2 \overline{u}_i}{\partial x_j\partial x_k} - 2\mu\overline{\dfrac{\partial u'_i}{\partial x_k}\dfrac{\partial u'_j}{\partial x_k}\dfrac{\partial u'_i}{\partial x_j}}}_{P_\varepsilon}$$

$$-2v\overline{\frac{\partial u_i'}{\partial x_k}\frac{\partial^2 p'}{\partial x_k \partial x_i}} - \underbrace{\overline{\mu u_j'\left(u_{i,k}'u_{i,k}'\right)_{,j}}}_{D_\varepsilon} - \underbrace{2\mu v\overline{\frac{\partial^2 u_i'}{\partial x_j \partial x_k}\frac{\partial^2 u_i'}{\partial x_j \partial x_k}}}_{\Phi_\varepsilon} + \mu\frac{\partial^2 \varepsilon}{\partial x_j \partial x_j} \quad (A30)$$

This equation is the exact ε-equation, which recovers equation (2.19). See Chapter 2 for explanations of P_ε, D_ε, and Φ_ε.

Further, derivation for ω-equation (2.29) can be performed using the exact equation for kinetic equation (A26) and ε-equation, using equation (2.25).

REYNOLDS AVERAGED EQUATION FOR SCALAR TRANSPORT

Transports of scalar quantities (e.g. temperature, mass) in turbulent flows can be, in general, categorized into two types: active and passive. An example of and active type is combustion. For passive type, we may consider, for example, turbulent non-isothermal flows. In this section, we derive the relevant equation for temperature Θ. The heat transfer equation reads

$$\frac{\partial \Theta}{\partial t} + (u_i \Theta)_{,i} = \alpha\Theta_{,ii} \quad (A31)$$

where $\alpha = K/\rho c_p$ is thermal diffusivity, K thermal conductivity, ρ density, and c_p is specific heat. After substituting for $\Theta = \overline{\Theta} + \Theta'$, as well as for velocity $u_i = \overline{u}_i + u_i'$, and averaging, we get

$$\frac{\partial \overline{\Theta}}{\partial t} + (\overline{u}_i \overline{\Theta})_{,i} = \alpha\overline{\Theta}_{,ii} - \left(\overline{u_i'\Theta'}\right)_{,i} \quad (A32)$$

Note the new term on the R.H.S. ($\overline{u_i'\Theta'}$). This term represents the turbulent heat flux, so-called Reynolds heat flux ($\rho c_p\overline{u_i'\Theta'}$), or mean correlation of fluctuating velocity and temperature. Similar to modeling Reynolds stress term in RANS, several models are available for turbulent heat flux term (i.e. a *closure*). One of these models is expressed as $\overline{u_i'\Theta'} = -\Gamma_T\overline{\Theta}_{,i}$ for which the constant Γ_T is turbulent diffusivity- a flow-dependent property. Analogous to laminar Prandtl number, a turbulent Prandtl/Schmidt number σ_T, is used to relate the turbulent momentum viscosity with turbulent diffusivity as $\Gamma_T = \dfrac{v_T}{\sigma_T}$, where it is usually assumed (and experimentally shown) that the value of σ_T is close to unity (between 0.7 and 1) which implies that the same turbulent mechanisms carry momentum and heat (i.e. $\Gamma_T \cong v_T$). Bachelor [97] suggested a diffusivity tensor, D_{ij} to be considered for an-isotropic turbulence, as $\overline{u_i'\Theta'} = -D_{ij}\overline{\Theta}_{,i}$. Equation (A32), including using a model for the turbulent heat flux, and a turbulence model are governing equations for non-isothermal turbulent flows. For further detail see the COMSOL manual (turbulent non-isothermal flow theory).

REFERENCES

1. J. Boussinesq, "Essai sur la théorie des eaux courantes," *Mémoires présentés par divers savants à l'Académie des Sciences*, vol. 23, no. 1, pp. 1–680, 1877.
2. O. Reynolds, "On the Dynamical Theory of Incompressible Viscous Fluids and Determination of the Criterion", vol. 186, pp. 123–164, 1895.
3. F. G. Schmitt, "About Boussinesq's turbulent viscosity hypothesis: historical remarks and a direct evaluation of its validity," *Comptes Rendus Mécanique*, vol. 335, no. 9–10, pp. 617–627, 2007.
4. Moin, P. and Kim, J., "Tracking Turbulence with Supercomputers," *Scientific American*, vol. 276, 1997.
5. H. Lamb, *Hydrodynamics*, Cambridge University Press, 1916.
6. M. Van Dyke, "An Album of Fluid Motion", Stanford, CA: The Parabolic Press, 1982.
7. NCFMF. [Online]. Available: http://web.mit.edu/hml/ncfmf.html. [Accessed 6 Feb. 2015].
8. R. Stewart, Director, *Turbulence.* [Film]. NCFMF, 1961.
9. D. o. F. Mechanics, "American Physical Society," [Online]. Available: *http://www.aps.org/units/dfd/pressroom/gallery/index.cfm*.
10. Samimy, M. et. al., A Gallery of Fluid Motion, Cambridge University Press, 2004.
11. Y. Li et. al., "A public turbulence database cluster and applications to study Lagrangian evolution of velocity increments in turbulence," *J. Turbulence*, vol. 9, no. 31, 2008.
12. F. M. White, *Fluid Mechanics, Seventh Edition*, McGraw-Hill, 2010.
13. T. Cebeci, *Analysis of Turbulent Flows, Third Edition*, Butterworth-Heinemann, 2013.
14. A. Tsinober, *An Informal Conceptual Introduction to Turbulence*, Springer Netherlands, 2009.
15. Reference has been intentionally deleted by the author
16. O. G. Bakunin, "Turbulence and Diffusion:Scaling Versus Equations", Springer, 2008.
17. Reference has been intentionally deleted by the author
18. L. Richardson, "Weather Prediction by Numerical Process", London: Cambridge University Press, 1922.

19. McDonough, "Introductory Lectures on Turbulence," 2007. [Online]. Available: *http://www.engr.uky.edu/~acfd/lctr-notes634.pdf*.

20. D. C. Wilcox, *Turbulence Modeling for CFD, Third Edition*, La Cañada, CA: DCW Industries, Inc., 2006.

21. Z. Warsi, Fluid Dynamics: Theoretical and Computational Approaches, Third Edition, CRC Press, 2005.

22. Tennekes, H., and Lumley, J.L, A First Course in Turbulence, MIT Press, 1972.

23. A. N. Kolmogorov, "The local structure of turbulence in incompressible viscous fluid for very large Reynolds number," *Proceedings of USSR Academy of Sciences,* vol. 30, pp. 299–303, 1941.

24. U. Frisch, "Turbulence, the Legacy of A. N. Kolmogorov", Cambridge: Cambridge University Press, 1995.

25. G. Batchelor, "The Theory of Homogeneous Turbulence", Cambridge: Cambridge University Press, 1953.

26. J. O. Hinze, *Turbulence, Second Edition*, New York: McGraw-Hill, 1975.

27. A. Townsend, "The Structure of Turbulent Shear Flow, Second Edition," Cambridge: Cambridge University Press 2014.

28. J. Westerweel, C. Fukushima, J.M. Pedersen, and J.C.R. Hunt, "Mechanics pf the Turbulent-Nonturbulent Interface of a Jet," *The American Physical Society,* vol. 95, pp. 1–4, 2005.

29. Chapman, G.T and Tobak, M., "Observations, Theoretical Ideas, and Modeling of Turbulent Flows," in *Theoretical Approaches to Turbulence*, D. e. al., Ed., New Yory, Springer-Verlag, 1985, pp. 19–49.

30. Lumley, J.L. and Yaglom, A.M., "A Century of Turbulence," *Flow, Turbulence and Combustion,* vol. 66, pp. 241–286, 2001.

31. Davidson, Peter A., et. al. (Editors), A Voyage Through Turbulence, Cambridge University Press, 2011.

32. H. Poncare, Les Methodes Nouvelles de la Mechanique Celeste, Paris: Gauthier Villars, 1899.

33. E. N. Lorenz, "Deterministic nonperiodic flow," *J. Atmos. Sci.,* vol. 20, pp. 130–141, 1963.

34. Orszag, S.A. and Patterson, G.S., "Numerical simulation of turbulence: statistical models and turbulence," *Lecture Notes in Physics,* vol. 12, pp. 127–142, 1972.

35. Schubauer, G. B. and Skramstad, H. K., "Laminar Boundary-Layer Oscillations and Transition on a Flat Plate," *NACA Rep.,* vol. 909, 1948.

36. P. Spalart, "Strategies for Turbulence Modeling and Simulations,"

International Journal of Heat and Fluid Flow, vol. 21, 2000.

37. Tarek Echekki, Epaminondas Mastorakos, Turbulent Combustion Modeling: Advances, New Trends and Perspectives, London: Springer, 2013.

38. J. Smagorinsky, "General circulation experiments with the primitive equations. I: The basic experiment," *Month. Weath. Rev.,* vol. 91, pp. 99–165, 1963.

39. Moin, P. and Mahesh, K., "Direct Numerical Simulation: A Tool in Turbulence Research," vol. 30, 1998.

40. Versteeg, H.K. and Malalasekera, W., "An Introduction TP Computational Fluid Dynamics-The Finite Volume Method", Pearson, Prentice Hall, Second Edition, 2007.

41. D. A. Fleisch, A Student's Guide to Vectors and Tensors, Cambridge: Cambridge University Press, 2012.

42. Schlichting, H., Gersten, K., Boundary-Layer Theory, 8th ed., Springer, 2000.

43. M. M. N. a. M. R. Lee, "Petascale direct numerical simulation of turbulent channel flow on up to 786K cores," in *International Conference on High Performance Computing, Network, Storage and Analysis,* 2013.

44. Bradshaw P., Cebeci T., and Whitelaw J.H., Engineering Calculation Methods for Turbulent Flow, London: Academic Press, 1981.

45. Anderson, D.A., Tannehill, J.C. and Pletcher, R.H., Computational Fluid Mechanics and Heat Transfer, New York: Hemisphere, 1984.

46. W. K. George, "Lectures in Turbulence for the 21st Century," 2013. [Online]. Available: *www.turbulence-online.com.*

47. Carati,D and Cabot,W., "Anisotropic eddy viscosity models," in *Centrer for Turbulence Research,* Stanford, CA, Proceedings of the Summer Program-1996.

48. L. Prandtl, "Über die ausgebildete Turbulez," Z. *Angew. Math. Mech.,* vol. 5, p. 136, 1925.

49. K. Hanjalic, "Advanced turbulence closure models: a view of current status and future prospects," *Int. J. Heat and Fluid Flow,* vol. 15, no. 3, pp. 178–203, 1994.

50. G. Alfonsi, "Reynolds-Averaged Navier-Stokes Equations for Turbulence Modeling," *Applied Mechanics Reviews,* vol. 62, no. 4, pp. 1–20, 2009.

51. Cebeci,T., and Smith,A.M.O., Analysis of Turbulent Boundary Layers, New York: Academic Press, 1974.

52. Baldwin,B.S and Lomax,H., "Thin Layer Approximation and Algebraic Model for Separated Turbulent Flows," *AIAA-78-0257,* Jan. 1978.

53. Agonafer, D., Gan-Li, L., and Spalding,D.B., "LVEL turbulence model for conjugate heat transfer at low Reynolds numbers," in *EEP 6, ASME International Mechanical Congress and Exposition*, Altanta, 1996.

54. W. Rodi, Turbulence Models and Their Application in Hyraulics- A state-of-the-art review, Rotterdam, Netherlands: IAHR, 1993.

55. *COMSOL manual>CFD Module>Single-Phase Flow>Theory for the Turbulent Flow User Interfaces*, COMSOL.

56. W. RODI, "Experience with Two-layer Models Combining the k-ε Model with a One-equation Model near the Wall," *AIAA*, pp. Papaer 91-0216, 1991.

57. Yakhot, V., Orszag, S.A., "Development of Turbulence Models for Shear Flows by a Double Expansion Technique," *Phys. Fluids A,* vol. 4, no. 7, pp. 1510–1520, 1992.

58. Patel V.C., Rodi W., Scheuerer G., "Turbulence Models for Near-Wall and Low Reynolds Number Flows: A Review," *AIAA J.*, vol. 23, pp. 1308–1319, 1985.

59. U. Schumann, "Realizability of Reynolds-stress turbulence models," *Physics of Fluids,* vol. 20, pp. 721–725, 1977.

60. Driver, D.M. and Seegmille, H.L., "Features of a Reattaching Turbulent Shear Layer in Diverging Channel Flow," *AIAA Journal,* vol. 23, pp. 163–171, 1985.

61. Bardina,J.E., Huang,P.G, Coakley,T.J., "Turbulence Modeling Validation, Testing, and Development," NASA, Ames Research Center, Moffett Field, CA, 1997.

62. NASA, "Menter Shear Stress Transport Model," Langley Research Center, Turbulence Modeling Resource, [Online]. Available: http://turbmodels.larc.nasa.gov/sst.html.

63. F. Menter, "Two-Equation Eddy-Viscosity Turbulence Models for Engineering Applications," *AIAA Journal,* vol. 32, no. 8, pp. 1598–1605, 1994.

64. Menter, F.R., Kuntz, M. and Langtry, R., "Ten Years of Industrial Experience with the SST Turbulence Model," *Turbulence Heat and Mass Transfer,* vol. 4, 2003.

65. F. Menter, "Zonal Two Equation k-ω Turbulent Models for Aerodynamic Flows," in *AIAA 24th Fluid Dynamics Conference*, 1993.

66. P. Bradshaw, *Turbulence, Second Edition*, Berlin: Springer-Verlag, 1978.

67. L. Prandtl, "The Mechanics of viscous fluids," in *Aerodynamics Theory, W.F. Durand, ed.*, Berlin, Julius Springer, 1935, pp. 34–208.

68. D. Spalding, "A single formula for the law of the wall," *Transaction of ASME, Serries E: Journal of Applied Mechanics,* vol. 28, pp. 455–458, 1961.

69. W. K. George, "Is there a universal log law for turbulent wall-bounded flows?," *Phil. Trans. R. Soc. A,* vol. 365, pp. 789-806, 2007.

70. H.Schlichting, K.Gersten, Boundary-Layer Theory, 8th ed., Springer, 2000.

71. Kim, S.-E. and Choudhury, D., "A near-wall treatment using wall functions sensitized to pressure gradient," *ASME FED Separated and Complex Flows,* vol. 217, pp. 273–279, 1995.

72. Abe, K., Kondoh, T. and Nagano, Y., "A New Turbulence Model for Predicting Fluid Flow and Heat Transfer in Separating and Reattaching Flows—I. Flow Field Calculations," *Int. J. Heat and Mass Transfer,* vol. 37, no. 1, pp. 139–151, 1994.

73. Chang, K.C., Hsieh, W.D. & Chen, C.S., "A modified low-Reynolds-number turbulence model applicable to recirculating flow in pipe expansion, Journal of Fluid Engineering," *Journal of Fluid Engineering,* vol. 117, pp. 417–423, 1995.

74. Launder, B. and Sharma, B., "Application of the energy dissipation model of turbulence to the calculation of flow near a spinning disc," *Lett. Heat and Mass transfer,* vol. 1, pp. 131–138, 1974.

75. Yang, Z., and Shih, T.H., "New time scale based k-ε model for near-wall turbulence," *1993,* vol. 31, no. 7, pp. 1191–1198, 1993.

76. Patel,V.C., Rodi, W., and Scheuerer, G., "Turbulence models for near-wall and low-Reynolds number flows: A Review," *AIAA,* vol. 23, p. 1308, 1985.

77. Jagadeesh, P. and Murali, K., "Application of Low-Re Turbulence Models for Flow Simulations Past Underwater Vehicle Hull Forms," *Journal of Naval Architecture and Marine Engineering,* vol. 1, pp. 41–54, 2005.

78. L. Davidson, "Chalmers University of Technology, Dept. of Thermo and Fluid Dynamics," January 2003. [Online]. Available: *http://www.fem.unicamp.br/~im450/palestras&artigos/kompendium_turb.pdf.*

79. Spalart, P. R. and Allmaras, S. R., "A One-EquationTurbulenceModel," *AIAA Paper 92-0439,* Jan. 1992.

80. N. L. R. Center, "Turbulence Modeling Resource," Jan. 2015. [Online]. Available: http://turbmodels.larc.nasa.gov/spalart.html. [Accessed Feb. 2015].

81. M. Tabatabaian, *"COMSOL for Engineers,"* Mercury Learning and Information, Dulles, VA, 2014.

82. Durst, F. and Loy, T., "Investigations of Laminar Flow in a Pipe with Sudden Contraction of Cross Sectional Area," *Computers & Fluids,* vol. 13, no. 1, pp. 15–36, 1985.

83. J. W. Slater, "NPARC Alliance CFD Verification and Validation," NASA, 1993. [Online]. Available: *http://www.grc.nasa.gov/WWW/wind/valid/validation.html.* [Accessed 18 05 2014].

84. A. G-077-1998, "Guide for the Verification and Validation of Computational Fluid Dynamics Simulations," AIAA, 1998.

85. Buice, C.U. and Eaton, J.K, "Experimental Investigation of Flow Through an Asymmetric Plane Diffuser," *Journal of Fluids Engineering,* vol. 122, pp. 433–435, 2000.

86. Samy M. El-Behery and Mofreh H. Hamed, "A Comparative Study of Turbulent Models Performance for Turbulent Flow in a Planar Asymmetric Diffuser," *World Academy of Science, Engineering and Technology,* vol. 53, pp. 769–780, 2009.

87. "NWTC Information Portal," https://wind.nrel.gov/airfoils/Airfoil-Families.html, 6 July 2012. [Online]. Available: https://wind.nrel.gov/airfoils/AirfoilFamilies.html. [Accessed 1 August 2014].

88. D. Somers, "Design and Experimental Results for S809 Airfoil," Airfoils, Inc., State College, PA, 1989.

89. "National Renewable Energy Laboratory," NREL, 21 October 2009. [Online]. Available: http://wind.nrel.gov/airfoils/shapes/S809_Shape.html. [Accessed 30 July 2014].

90. Wolfe, Walter P., Ochs, Stuart S., "CFD Calculations of S809 Aerodyamic Characteristics," *AIAA-97-0973,* pp. 1–8.

91. W. Dean, "Note on the motion of fluid in a curved pipe," *Phil. Mag.,* vol. 4, pp. 208–223, 1927.

92. G. Homicz, "Computational Fluid Dynamics Simulations of Pipe Elbow Flow," Sandia National Laboratories, Albuquerque, NM, 2004.

93. R. Blevins, Applied Fluid Dynamics Handbook, New York: Van Nostrand Reihold Co., 1984.

94. Mazoni, D and Guitton, P., "Validation of Displacement Ventilation Simplified," in *the Fifth International IBSPA Conference, September 8–10, 1997, Volume I, pp 233–239*, Prague, Czech Republic, 1997.

95. Rodi, W., Editor, "Comparison of Turbulence Models in Case of Jet in Crossflow using Commercial CFD Code," in *Engineering Turbulence Modelling and Experiments 6: ERCOFTAC International Symposium on Engineering Turbulence and Measurements - ETMM6,* Elsevier, 2005, pp. 399–408.

96. Ozcan, O. and Larsen, P.S, "An Experimental study of a turbulent jet in cross-flow by using LDA," Technical University of Denmark, Report MEK-FM 2001–02.

97. G. Batchelor, "Diffusion in a field of homogeneous turbulence," *Aust. J. Sci. Res.,* vol. 2, no. A, pp. 437–450, 1949.

98. N. C. f. F. M. Films, "Illustrated Experiments in Fluid Mechanics-The NCFMF Book of Film Notes", A. H. Shapiro, Ed., MIT Press, 1984.

99. T. V. Karman, "Some remarks on the statistical theory of turbulence," in *Proc. 5th Int. Congr. Appl. Mech.*, Cambridge, MA, 1938.

100. L. F. Richardson, Weather Prediction by Numerical Process, Cambridge University Press, 1922.

101. N.I.Muskhelishvili, Some Basic Problems of the Mathematical Theory of Elasticity, Leyden: Noordhoff Internation Publishing, 1954.

102. O. Reynolds, "On the experimetal investigation of the circumstances which determine wether the motion of water shall be direct or sinuous and the law of resistance in in parallel channels," *Phil. Trans. Roy. Soc.*, vol. 174, p. 935, 1883.

103. S. Pope, Turbulent Flows, Cambridge University Press, 2000.

104. Loiola,B.R., Altemani,C.A.C., "Experimental Evaluation of the Covective and the Conjugate Cooling of a Protruding Heater in a Duct," in *Proceedings of the 14th Brazilian Congress of Thermal Sciences and Engineering*, Rio de Janeiro, Brazil, 2012.

105. Loiola, Bruna R., Altemani, Carlos A. C., "Comparative Numerical and Experimental Results for the Conjugate Cooling of a Discrete Heater in a Duct," in *22nd International Congress of Mechanical Engineering*, Ribeirão Preto, SP, Brazil, 2013.

INDEX